T0224352

Introduction to R for Terrestrial Ecology

Milena Lakicevic • Nicholas Povak
Keith M. Reynolds

Introduction to R for Terrestrial Ecology

Basics of Numerical Analysis,
Mapping, Statistical Tests
and Advanced Application of R

 Springer

Milena Lakicevic
Faculty of Agriculture
University of Novi Sad
Novi Sad, Serbia

Nicholas Povak
Pacific Northwest Research Station
United States Department of Agriculture
Wenatchee, WA, USA

Keith M. Reynolds
Pacific Northwest Research Station
USDA Forest Service
Corvallis, OR, USA

ISBN 978-3-030-27605-8   ISBN 978-3-030-27603-4   (eBook)
https://doi.org/10.1007/978-3-030-27603-4

This Springer imprint is published by the registered company Springer Nature Switzerland AG
The registered company address is: Gewerbestrasse 11, 6330 Cham, Switzerland

Acknowledgments

Part of this research was conducted in the USDA Forest Service in Corvallis, Oregon, USA, during a 6-month visit of the first author, sponsored by the Fulbright program of the U.S. Department of State.

Also, we thank our reviewers: Dr. María Jacoba Salinas (University of Almería, Spain), Dr. María Mercedes Carón (Instituto Multidisciplinario de Biología Vegetal, Argentina), Dr. Neşe Yılmaz (Istanbul University, Turkey), and Dr. Keivan Shariatmadar (KU Leuven University Belgium, Belgium) for their detailed comments and suggestions that helped to improve the quality of the final version of this book.

About the Book

This textbook covers R data analysis related to terrestrial ecology, starting with basic examples and proceeding up to advanced applications of the R programming language. The main objective of the textbook is to serve as a guide for undergraduate students, who have no previous experience with R, but Chap. 5 is dedicated to advanced R applications and will also be useful for master's and PhD students and professionals.

The textbook deals with solving specific programming tasks in R, and tasks are organized in terms of gradually increasing R proficiency, with examples getting more challenging as the chapters progress. The main competencies students will acquire from this textbook are:

- Manipulating and processing data tables
- Performing statistical tests
- Creating maps in R
- Advanced R applications

All examples in the book are solved using R and its interface RStudio. The installation process is quite easy; firstly, R should be downloaded from https://cran.r-project.org/ and then RStudio is to be downloaded from https://www.rstudio.com/products/rstudio/download/ following very few installation instructions.

Some sections refer to importing data sets from Excel files for use with working through steps of an analytical process. Appendices 1, 2, and 3 provide previews of these Excel files, and readers can download the Excel files from the Springer page at https://www.springer.com/us/book/9783030276027. When unzipped, you will have a folder named IntroToRExcelFiles that can be placed anywhere on your hard drive.

This textbook will be useful in undergraduate and graduate courses in advanced landscape ecology, analysis of ecological and environmental data, ecological modeling, analytical methods for ecologists, statistical inference for applied research, elements of statistical methods, computational ecology, landscape metrics, and spatial statistics.

Apart from its application in ecology, this book can be useful for other scientific fields, especially for biology, forestry, and agriculture. Tasks presented in this book can be solved by applying other programs and computer tools, but our intent was to promote application of R and to demonstrate its main strengths and possibilities.

Contents

About the Authors

Milena Lakicevic is an Associate Professor at the University of Novi Sad, Faculty of Agriculture in Novi Sad, Serbia. She holds a PhD in biotechnical sciences and her primary area of expertise is application of R programming, multi-criteria analysis, and decision support systems in landscape and natural resources management. Dr. Lakicevic spent 6 months (January–July 2018) at the Pacific Northwest Research Station (U.S. Department of Agriculture, Forest Service) as a Fulbright Visiting Scholar, working on spatial decision support systems under the supervision of Dr. Keith Reynolds. She has published over 60 scientific papers and served as a reviewer for international journals and has given guest lectures at several universities in Europe.

Nicholas Povak is a Postdoctoral Fellow with the Oak Ridge Associated Universities and the Pacific Northwest Research Station (US Forest Service). He received his PhD from the University of Washington—Seattle, where he explored the application of machine learning algorithms for the prediction of ecological processes such as stream water quality and cavity nesting bird habitat. Dr. Povak worked as a postdoc with the Institute of Pacific Islands Forestry (Hilo, HI) developing decision support models to help manage invasive species, hydrologic flows, and sediment yields in tropical forested watersheds. Currently, Dr. Povak is applying machine learning algorithms in the R programming environment to predict fire severity patterns and controls on fire spread from satellite data and to better understand biophysical drivers of post-fire conifer regeneration. He also creates web applications using the Shiny R package to provide a portal for stakeholders to interact with decision support models to aid in their decision-making process.

Keith M. Reynolds is a research forester with the Pacific Northwest Research Station (U.S. Department of Agriculture, Forest Service) and is located at the Corvallis Forestry Sciences Laboratory in Corvallis, OR. Although Dr. Reynolds' advanced degrees are in quantitative plant epidemiology, his primary areas of expertise are in statistics, biomathematics, and knowledge-based systems theory and application. He has been the team leader of the Ecosystem Management Decision Support project at the PNW Station since 1993, designing and implementing new spatially enabled knowledge-based systems technologies for environmental analysis and planning. EMDS version 1.0 was released in February 1997, and the system is now at version 7.1. For the past 20 years, Dr. Reynolds has also been working in a wide variety of specific application areas, including watershed assessment, forest ecosystem sustainability, fish and wildlife habitat suitability, regional planning, national forest-fuels management, landscape restoration, environmental effects of atmospheric sulfur deposition, maintenance of national environmental infrastructure for the Army Corps of Engineers, and the national Terrestrial Condition Assessment of the US Forest Service. He has published over 125 refereed articles and book chapters and has edited five volumes, including *Making Transparent Environmental Management Decisions: Applications of the Ecosystem Management Decision Support System* (Springer 2014).

List of Figures

List of Tables

Types of Data in R

In this chapter, we explain how to create and work with the main types of data in R: vectors, matrices, and data frames. In addition, we briefly explain how to create lists and merge them with data frames. At the end of the chapter, we discuss how to deal with missing data and outliers in data sets.

1.1   Vectors

A vector is a string of elements of the same type. There are four types of vectors:

1. numerical
2. character (string)
3. logical
4. factor

We discuss each of them separately and explain how to perform some basic operations on them.

1.1.1   Numerical Vectors

First, we explain how to create a numerical vector, and then how to operate on data within the vector.

1.1.1.1 Creating a Numerical Vector

We can create a numerical vector named num.vector that contains the numbers: 5, 12, 25, 3, 4, and 10, with the following command:

```
#Creating a numerical vector
num.vector<-c(5, 12, 25, 3, 4, 10)
num.vector
```

And as a result, R replies with:

[1] 5 12 25 3 4 10.

© Springer Nature Switzerland AG 2020
M. Lakicevic et al., *Introduction to R for Terrestrial Ecology*,
https://doi.org/10.1007/978-3-030-27603-4_1

The generic function c() is used to concatenate values together into a vector. It is required when combining individual values or a series of numbers and vectors together as shown in subsequent examples. Other functions can be used to explore vector properties, including the class, vector size, and quantiles.

```
#Exploring a numerical vector
class(num.vector)
```

[1] "integer"

```
#Exploring a numerical vector
length(num.vector)
```

[1] 6

```
#Summary function for a numerical vector
summary(num.vector)
```

```
  Min.  1st Qu.  Median  Mean  3rd Qu.    Max.
 3.000   4.250   7.500  9.833  11.500   25.000
```

Notice the conventions that c() is a function whose arguments specify the contents of the vector (with elements separated by commas in this case), <- indicates an assignment operation, and the newly created vector in this case is named **num.vector**.

Also note that the class "integer" is a special case of class "numeric," in which all values are whole numbers.

Now, we can consider some other ways of defining the contents of a vector. For example, if we need a vector containing numbers from 10 to 20, we do not need to type them all; instead we can do the following:

```
#Creating a numerical vector – consecutive numbers (10 to 20)
ten.to.twenty<-10:20
ten.to.twenty
```

This way, we get:

[1] 10 11 12 13 14 15 16 17 18 19 20

Here, the colon operator generates a vector of a regular sequence of integers from 10 – 20, and the c() function is not required. However, placing c()around 10:20 will not affect the result, nor cause an error.

There also are other options. For example, to create a vector containing numbers from 1 to 5, the number 10, and numbers from 15 to 20, we use the following command:

```
#Creating a numerical vector (from 1 to 5,  10,  from 15 to 20)
num.string.1<-c(1:5, 10, 15:20)
num.string.1
```

And we get:

[1] 1 2 3 4 5 10 15 16 17 18 19 20

Also, to create a numerical vector that contains the elements (0,1,0,1,0,1,0,1, 0,1), notice that we have a pair (0,1) repeating five times. Therefore, we can apply the function "rep":

```
#Creating a numerical vector – repeating the (0,1) pair 5 times
num.string.2<-rep(0:1, 5)
num.string.2
```

And we get the expected outcome:

[1] 0 1 0 1 0 1 0 1 0 1

Again, the function rep() results in a vector and the function c() is not required. As another example, to create a vector that contains five zeros and four ones (0,0,0,0,0,1,1,1,1), we can use the "rep" function to create a string of elements like this:

```
#Creating a numerical vector – repeating values in a string
num.string.3<-c(rep(0,each=5), rep(1, each=4))
num.string.3
```

As a result, R returns:

[1] 0 0 0 0 0 1 1 1 1

Here, c() is required because we are combining two vectors into a single vector. If we do not include c(), R will throw the following error:

```
Error: unexpected ',' in "num.string.3 <- rep(0,each=5),"
```

As we have demonstrated in the previous examples, using syntax with ":" and/or function "rep" makes creating numerical vectors fast and easy. There are also some other possibilities for selecting, excluding, and adding elements, and these actions can be performed on all types of vectors (numerical, string, and logical), so they are explained at the end of this chapter.

1.1.1.2 Manipulating a Numerical Vector

Once we have a created a vector, we can apply different functions to it. We demonstrate some basic functions on the vector named **num.vector**, which we have already created. Here are some examples:

```
#Finding the maximum value
max(num.vector)

[1] 25

#Finding the minimum value
min(num.vector)

[1] 3

#Counting the number of elements
length(num.vector)

[1] 6
```

```
#Calculating the sum of all elements
sum(num.vector)
```

[1] 59

```
#Calculating the average value
mean(num.vector)
```

[1] 9.833333

```
#Calculating the average value and rounding on two decimal places
round(mean(num.vector),2)
```

[1] 9.83

```
#Calculating the median value
median(num.vector)
```

[1] 7.5

```
#Putting the vector elements in reverse order
rev(num.vector)
```

[1] 10 4 3 25 12 5

```
#Sorting the vector elements in ascending order
sort(num.vector)
```

[1] 3 4 5 10 12 25

```
#Sorting the vector elements in descending order
sort(num.vector, decreasing = T)
```

[1] 25 12 10 5 4 3

We can also create new objects to hold the results of the above operations for later use:

```
#Create new object for the maximum of the vector
max.vector <- max(num.vector)
max.vector
```

[1] 25

1.1.2   Character Vectors

Besides numerical vectors, we often need some character or string elements to be included in our research. In R, the terms character and string are interchangeable in theory, but in practice the specific class is referred to as character. In environmental sciences, we commonly need to specify properties such as name of plant or animal species, names of locations, type of habitat, etc. To create a character vector, we use the same functions as for a numerical vector, but in this case the elements of a vector are separated with the sign "". For example, to create a character or string vector consisting of the following elements: Species 1, Species 2, Species 3, Species 4, Species 5, we apply the following command:

```
#Creating a character vector
char.vector<-c("Species 1", "Species 2", "Species 3",
"Species 4", "Species 5")
char.vector
```

And as a result, we get:

[1] "Species 1" "Species 2" "Species 3" "Species 4" "Species 5"

For this example, there is also a shortcut for creating this particular string vector. Because we have the same word repeating (Species) followed by consecutive numbers from 1 to 5, we can apply the following command:

```
#Creating a vector of strings (using the "paste" function)
char.vector.p<-paste("Species", 1:5)
char.vector.p
```

And once again, we get the result:

"Species 1" "Species 2" "Species 3" "Species 4" "Species 5"

We can check that the vectors named **char.vector** and **char.vector.p** are indeed the same with the function "==":

```
#Checking if the elements of two vectors are the same
char.vector==char.vector.p
```

And as a result, we get:

[1] TRUE TRUE TRUE TRUE TRUE

Meaning that all elements of these two vectors are exactly the same.

Another option would be to check whether any elements of the vectors **char.vector** and **char.vector.p** are different, and that can be done by applying the function "!=":

```
#Checking if the elements of two vectors are different
char.vector!=char.vector.p
```

And this time we get the result:
[1] FALSE FALSE FALSE FALSE FALSE

Meaning that there is no difference between elements of these two vectors.

A final option to evaluate whether all elements within two vectors are equal:

```
#Checking if all elements of two vectors are the same
all.equal(char.vector, char.vector.p)
```

[1] TRUE

Note that both the function "==" and the function "!=" can be applied to all types of vectors, and these functions are convenient for making comparisons between elements.

Also, we can create a string of elements that are repeating, for example, if we want to create a string vector with the following elements: habitat A, habitat B, habitat B, habitat C, habitat C, habitat C, we basically need to create a string with habitat A repeating once, habitat B repeating twice and habitat C repeating three times.

```
#Creating a character vector with repeating elements
habitat<-c("habitat A", rep("habitat B", each=2),
rep("habitat C", each=3))
habitat
```

And we get:

```
[1] "habitat A" "habitat B" "habitat B" "habitat C" "habitat C" "habitat C"
```

Another useful function is factoring, which returns a list of all the unique elements of a vector. For example, the previous vector can be factored as:

```
#Creating a factor
habitat.factor<-factor(habitat)
habitat.factor
```

And we get:

```
[1] habitat A habitat B habitat B habitat C habitat C habitat C
Levels: habitat A habitat B habitat C
```

In this way, we have defined different levels for the vector named **habitat.factor**. Now we can apply the "summary" function:

```
#Summary function for a factor
summary(habitat.factor)
```

And we get the result:

```
habitat A habitat B habitat C
        1         2         3
```

The function "factor" is useful for grouping the elements in cases when we have different levels, categories, etc.

Lastly, we check what happens if we have combined strings and numbers when creating a vector:

```
#Creating a character vector – turn numbers into characters
species<-c("Quercus rubra", "Tilia cordata", 100)
species
```

As a result, we get:

```
[1] "Quercus rubra" "Tilia cordata" "100"
```

And this means that the number 100 has been converted to the string data type. As already noted, vectors always contain elements of the same type, so we cannot combine strings with numbers without converting numbers into strings in the process.

To check the number of elements, that is easily done by:

```
#Counting number of elements in a character vector
length(species)
```

And as a result, we have:

[1] 3

1.1.3   Logical Vectors

Logical vectors consist of binary data: true and false. Creating a logical vector is very easy, and one example is:

```
#Creating a logical vector
logical.vector<-c(T,T,F,F,T,F)
logical.vector
```

And this way we get:

[1] TRUE TRUE FALSE FALSE TRUE FALSE

As in the previous examples, we can use the function "rep":

```
#Creating a logical vector – repeating the pair (T,F) 4 times
l1.vector<-rep(c(T,F), 4)
l1.vector

[1]  TRUE FALSE  TRUE FALSE  TRUE FALSE  TRUE FALSE

#Creating a logical vector – repeating T and F  4 times each
l2.vector<-c(rep(T, each=4), rep(F, each=4))
l2.vector

[1]  TRUE  TRUE  TRUE  TRUE FALSE FALSE FALSE FALSE
```

One interesting feature of the logical vector is that we can apply the function "sum." If we sum the elements in the logical vector named **l2.vector**:

```
#Summing the elements of a logical vector
sum(l2.vector)
```

we get the result:

[1] 4

The value of 4 has been calculated based on the rule that TRUE is equal to 1 and FALSE is equal to zero. This feature can be useful when analyzing data (for example, if we have a dataset with certain measurements and the associated column with "TRUE"/"FALSE" values, where "TRUE" means that data are verified, summing the values will show us the number of measurement that are verified, etc.).

1.1.4   Selecting, Removing, and Adding Elements with Vectors

In this section, we show how to select (keep) and remove elements from a vector and add elements to a vector. All the operations are shown on a vector that we create first:

```
#Creating a vector
vector<-c(10,15,100,32,64,50)
vector
```

1.1.4.1 Selecting Elements from a Vector

First, we show a procedure for selecting elements from a vector, based on their position in the vector. This can be done using "[n]" – square brackets and a number or numbers indicating the place of one or more elements in the vector. If we have sequential numbers, we can apply "$n_1:n_2$" or "$c(n_1,n_2)$" if the numbers are not consecutive.

```
#Selecting the second element
vector[2]

[1] 15

#Selecting the fifth and sixth elements
vector[5:6]

[1] 64 50

#Selecting the first, fourth and sixth elements
vector[c(1,4,6)]

[1] 10 32 50

#Selecting the first, second, fifth and sixth elements
vector[c(1:2, 5:6)]

[1] 10 15 64 50
```

Next, we explain how to select elements from a vector based on a certain condition. For example, we want to list elements from our vector that are greater than 25.

```
#Checking the condition (>25)
vector>25

[1] FALSE FALSE  TRUE  TRUE  TRUE  TRUE

#Selecting the elements fulfilling the condition (>25)
vector[vector>25]

[1] 100  32  64  50
```

Notice the difference between the above two expressions. The first expression simply returns true or false for each element in vector relative to the condition, >25. However, in the second expression, the brackets mean that we want a **selection** on the elements of vector, based on the condition that a selected element is >25.

We can combine two or more conditional expressions into a compound expression. For example, we can get a selection of all elements that are > 20 and < 60. Here, we use the operator "&:"

```
#Selecting the elements fulfilling two conditions
vector[vector>20 & vector<60]
```

[1] 32 50

Another example is to make a selection of elements that are < 20 or > 70. For this example, we need the operator "|" symbolizing "or:"

```
#Selecting the elements fulfilling two conditions
vector[vector<20 | vector>70]
```

[1] 10 15 100

Next, we show how to perform an operation using a certain condition. For example, let's find the sum of all values larger than 50:

```
#Calculating a sum for elements fulfilling a condition
sum(vector[vector>50])
```

[1] 164

As a final example, we show how to perform an operation using more than one condition. For example, let's find an average value for all elements greater than 10 and smaller than 50:

```
#Calculating an average for elements fulfilling two conditions
mean(vector[vector>10 & vector<50])
```

[1] 23.5

1.1.4.2 Removing Elements from a Vector

The syntax for removing elements from a vector is similar to the syntax for selecting them. The only difference is in the use of the minus sign, "-", preceding an element number or command that operates on one or more vector elements:

```
#Removing the second element
vector[-2]
```

[1] 10 100 32 64 50

```
#Removing the fifth and sixth elements
vector[-c(5:6)]
```

[1] 10 15 100 32

```
#Removing the first, fourth and sixth elements
vector[-c(1,4,6)]
```

[1] 15 100 64

```
#Removing the first, second, fifth and sixth elements
vector[-c(1:2, 5:6)]
```

[1] 100 32

An important point is that, unless we assign the object "vector" using either the "<-" operator, as used previously, or the "=" operator, the object "vector" will remain unchanged. For example, in the above examples, the original object "vector" does not change because we have not assigned it to an object:

```
#Removing the second element and assigning it to vector object
vector <- vector[-2]
vector

[1] 10 100  32  64  50
```

The object vector in this example went from length 6 to length 5 by removing the second element.

1.1.4.3 Adding Elements to a Vector

First, we show how to add elements at the end of the **vector**. Let's add numbers 10 and 70 at the end of the vector:

```
#Adding elements at the end of a vector
c(vector,10,70)

[1]  10 100  32  64  50  10  70
```

Similarly, to add the numbers 10 and 70 at the beginning of the **vector,** we write:

```
#Adding elements at the beginning of a vector
c(10,70,vector)

[1]  10  70  10 100  32  64  50
```

To insert the numbers 10 and 70 between the first and second element of **vector,** we write:

```
#Adding elements within a vector
c(vector[1], 10,70, vector[2:5])

[1]  10  10  70 100  32  64  50
```

Although we have demonstrated the procedures for selecting, removing, and adding elements to a numerical vector, the same syntax is equally applicable to character and logical vectors.

1.2   Matrices

Matrices are two-dimensional arrays consisting of rows and columns, and whose elements are all of the same type (e.g., numeric, string, or logical). First, we create a simple one-column matrix that will contain high air temperatures by month of the year for some location.

```
#Creating a matrix
high.temperature<-matrix(c(48,52,56,61,68,74,
82,82,77,64,53,46), ncol=1, byrow=T)

high.temperature
```

This way, we created a matrix with one column (that was defined by "ncol=1") and the values from the numerical vector were put in a matrix by filling the rows first (that was defined by "byrow=T"). Actually, in this example, we could have excluded the parameter "byrow=T," because we had just one column, so there was no other option for filling the matrix, but we have shown the general form of the matrix syntax for the general case when defining a matrix with more than one column.

If we want a matrix to be filled by rows first, then we apply the parameter "byrow=T", and in case we want a matrix to be filled by columns first, we can also specify the parameter "byrow=F".

In the previous step, we created a matrix without names for the columns and rows. To assign names to rows and columns, we do the following:

```
#Assigning names to rows and columns
rownames(high.temperature)<-c("Jan", "Feb", "March",
"April", "May", "June", "July",
"Aug", "Sep", "Oct", "Nov", "Dec")

colnames(high.temperature)<-c("high temperature [F]")

high.temperature
```

and the result would be:

	high temperature [F]
Jan	48
Feb	52
March	56
April	61
May	68
June	74
July	82
Aug	82
Sep	77
Oct	64
Nov	53
Dec	46

1.2.1 Manipulating a Matrix

Now, we show how to manipulate data within a matrix. For basic operations, the procedure is similar to the one for vectors.

```
#Calculating average value for high temperature
mean(high.temperature)
```

[1] 63.58333

```
#Calculating median value for high temperature
median(high.temperature)
```

[1] 62.5

We can also add new columns to the matrix. Let's say that we want to add a column with average low temperature at the same location.

```
#Creating a matrix with low temperature
low.temperature<-matrix(c(35,35,37,40,45,49,53,53,48,42,38,34),
ncol=1)
colnames(low.temperature)<-c("low temperature [F]")
low.temperature

#Adding a new column
matrix.1<-cbind(high.temperature, low.temperature)
matrix.1
```

Here is the result:

	high temperature [F]	low temperature [F]
Jan	48	35
Feb	52	35
March	56	37
April	61	40
May	68	45
June	74	49
July	82	53
Aug	82	53
Sep	77	48
Oct	64	42
Nov	53	38
Dec	46	34

The function "cbind" is used here to create a new matrix named matrix.1 whose columns are the row-wise combination of high.temperature and low.temperature.

Now we show how to manipulate data within a matrix. To calculate the difference between high and low average temperatures, we apply the following function:

```
#Subtracting columns of matrix
difference<-matrix.1[,1]-matrix.1[,2]
difference
```

In this function, we calculated the difference between the first column of the **matrix.1** (that is defined by "matrix.1[,1]") and the second column of the matrix.1 (that is defined by "matrix.1[,2]") and as a result we get:

Jan	Feb	March	April	May	June	July	Aug	Sep	Oct	Nov	Dec
13	17	19	21	23	25	29	29	29	22	15	12

We can add the column with the difference value to the matrix.1, and update the names of columns:

```
#Adding columns to a matrix
matrix.2<-cbind(matrix.1, difference)
colnames(matrix.2)<-c(colnames(matrix.1), "difference [F]")

matrix.2
```

And we get the following output:

	high temperature [F]	low temperature [F]	difference [F]
Jan	48	35	13
Feb	52	35	17
March	56	37	19
April	61	40	21
May	68	45	23
June	74	49	25
July	82	53	29
Aug	82	53	29
Sep	77	48	29
Oct	64	42	22
Nov	53	38	15
Dec	46	34	12

We can use the following syntax to select a single column or a single row and return a numeric vector:

```
#Select column 2
matrix.2[, 2]

Jan Feb March April May June July Aug Sep Oct Nov Dec
 35  35    37    40  45   49   53  53  48  42  38  34
#Select row 3
matrix.2[3, ]

high temperature [F]  low temperature [F]    difference [F]
                56                   37                 19

#Select June row
matrix.2[which(rownames(matrix.2) == "June"), ]

high temperature [F]  low temperature [F]    difference [F]
                74                   49                 25
```

To calculate the average value for each column, the easiest way is by using the function "apply." The general syntax of this function is "apply(data, 1, function)," if we are processing data in rows, or "apply(data, 2, function)," if we are dealing with data in columns. In our example, to find the average of each column, we use the second syntax (with number 2), and specify the name of the function, "mean":

```
#Calculating average of columns of a matrix
average<-apply(matrix.2,2, mean)
average<-round(average,1)
average
```

In this last step, we calculated the average value of each column, and rounded it to one decimal place. To add a new row to the **matrix.2** with the average values we just calculated, we apply the function "rbind":

```
#Adding a row to a matrix
rbind(matrix.2, average)
```

And we get the following result:

	high temperature [F]	low temperature [F]	difference [F]
Jan	48.0	35.0	13.0
Feb	52.0	35.0	17.0
March	56.0	37.0	19.0
April	61.0	40.0	21.0
May	68.0	45.0	23.0
June	74.0	49.0	25.0
July	82.0	53.0	29.0
Aug	82.0	53.0	29.0
Sep	77.0	48.0	29.0
Oct	64.0	42.0	22.0
Nov	53.0	38.0	15.0
Dec	46.0	34.0	12.0
average	63.6	42.4	21.2

Next, we show how to manipulate data from a portion of a matrix. Let's say we want to calculate both average high and average low temperatures for just the months of July, August, and September. We select the appropriate rows and columns from the matrix, **matrix.2,** and use the function "apply:"

```
#Manipulating data within a matrix
summer<-apply(matrix.2[c(7:9),1:2], 2, mean)
names(summer)<-c("summer high t [F]", "summer low t [F]")
summer<-round(summer,2)

summer
```

We get the result:

summer high t [F]	summer low t [F]
80.33	51.33

There are many other options for manipulating data within a matrix, but they are similar to the options for manipulating data within a data frame, so, in the next section, we provide additional examples that are also pertinent to matrices.

1.3   Data Frames and Lists

Data frames are two-dimensional arrays that can contain objects of different types. For example, a data frame can consist of numerical, string, and logical data vectors. This is the main difference between a matrix and a data frame; a matrix always contains the same type of elements (for example, numbers), but a data frame can include different types of data. For that reason, data frames are more commonly used in environmental research.

1.3.1 Creating a Data Frame

Let's say we want to create a data frame with the following structure: id, species name, size of population, altitude range, and protection status. This dataset would correspond to some inventory data we collected while doing some field work. If we analyze this dataset more closely, we can conclude that "id" could be a numerical vector, "species name" should be a string vector, "size of population" should be a numerical vector, "altitude range" is going to be created as a string vector, and "protection status" as logical, with a TRUE value indicating that this species is recognized as endangered and should be protected, and FALSE value indicating that the species is not assessed as endangered. Next, we demonstrate creating a data frame that contains all three types of vectors. First, we create each vector:

```
#Creating vectors
id<-1:12
species<-paste("Species", 1:12)
size<-c(50,25,30,45,2,70,22,20,10,45,22,56)
altitude<-c(rep("0-500",2), rep("501-1000",3), rep("1001-1500",5), rep("0-500",2))
protection<-c(rep(T, each=5), rep(F, each=7))
```

Now we have the data, and the next step is to create a data frame. A data frame is created by applying the function "data.frame(names of columns-vectors):"

```
#Creating a data frame
df<-data.frame(id, species, size, altitude, protection)
df
```

And we will get the data frame **df**:

id		species	size	altitude	protection
1	1	Species 1	50	0-500	TRUE
2	2	Species 2	25	0-500	TRUE
3	3	Species 3	30	501-1000	TRUE
4	4	Species 4	45	501-1000	TRUE
5	5	Species 5	2	501-1000	TRUE
6	6	Species 6	70	1001-1500	FALSE
7	7	Species 7	22	1001-1500	FALSE
8	8	Species 8	20	1001-1500	FALSE
9	9	Species 9	10	1001-1500	FALSE
10	10	Species 10	45	1001-1500	FALSE
11	11	Species 11	22	0-500	FALSE
12	12	Species 12	56	0-500	FALSE

1.3.2   Selecting Data Within Data Frames

Now, we look at how to select data from a data frame. For example, to select a certain element from the data frame, we use the command "data[nrow, ncol]". Let's say that we want to select an element from the second row and the third column. In that case, we apply:

```
#Selecting an element in a data frame
df[2,3]

[1] 25
```

If we want to extract the entire row, we use the command "data[nrow,]". For example, if we are selecting the fourth row, we apply:

```
#Selecting a row from a data frame
df[4,]

id    species  size   altitude   protection
4  4  Specie 4   45   501-1000         TRUE
```

It is also possible to select more than one row. To select the first, fifth, and sixth row, we apply:

```
#Selecting rows from a data frame
df[c(1,5:6),]

id    species   size   altitude  protection
1  1  Species 1   50     0-500         TRUE
5  5  Species 5    2   501-1000        TRUE
6  6  Species 6   70  1001-1500       FALSE
```

Selecting a column from a data frame is similar to selecting a row. In this case, we need to apply the command "data[,ncol]." Therefore, to extract the fifth column from a data frame, we apply:

```
#Selecting a column from a data frame
df[,5]

[1] TRUE  TRUE  TRUE  TRUE  TRUE FALSE FALSE FALSE FALSE FALSE
FALSE FALSE
```

An alternative option for selecting a column from a data frame uses the command "data$name. of.column". For example, to extract the column named "protection," we apply:

```
#Selecting a column from a data frame
df$protection
[1] TRUE  TRUE  TRUE  TRUE  TRUE FALSE FALSE FALSE FALSE FALSE
FALSE FALSE
```

To select, for example, the first three columns, we apply:

```
#Selecting columns from a data frame
df[,1:3]

id        species  size
1    1    Species 1   50
2    2    Species 2   25
3    3    Species 3   30
4    4    Species 4   45
5    5    Species 5    2
6    6    Species 6   70
7    7    Species 7   22
8    8    Species 8   20
9    9    Species 9   10
10  10   Species 10   45
11  11   Species 11   22
12  12   Species 12   56
```

It is also possible to define a condition for selecting both certain rows and certain columns at the same time. For example, to select rows 3, 4, and 6 and columns 2 and 3, we apply:

```
#Selecting rows and columns from a data frame
df[c(3:4,6),c(2:3)]

    species  size
3  Species 3   30
4  Species 4   45
6  Species 6   70
```

In addition, selecting columns and rows can be based on certain conditions related to our data. For example, we can select part of the data frame with size of population ≥ 50, and for that purpose we will use the following command "data[condition,]:"

```
#Selecting elements that fulfill certain condition
df[df$size>=50, ]
```

And we get:

```
    id    species  size   altitude  protection
1    1   Species 1   50     0-500       TRUE
6    6   Species 6   70  1001-1500      FALSE
12  12  Species 12   56     0-500       FALSE
```

We can also make a selection by using the function "subset." The general command is "subset(data, condition)." If we consider the same condition – selecting a part of data frame with population size is equal or greater than 50, we apply:

```
#Subset function - one condition
subset(df, size>=50)

id          species  size    altitude  protection
1    1    Species 1   50       0-500       TRUE
6    6    Species 6   70    1001-1500      FALSE
12  12   Species 12   56       0-500       FALSE
```

In addition, we can define multiple selection criteria. For example, we can select that part of the data frame for which population size is >10, and altitude is "0-500" or "501-1000," and protection status is TRUE.

```
#Subset function - several conditions
subset(df, size>10 &(altitude=="0-500"|altitude=="501-1000")
&protection==T)

id          species  size    altitude  protection
1    1    Species 1   50       0-500       TRUE
2    2    Species 2   25       0-500       TRUE
3    3    Species 3   30     501-1000      TRUE
4    4    Species 4   45     501-1000      TRUE
```

To fulfill these conditions, we used the "and" and "or" logical operators discussed previously. We used "and" between conditions for size, altitude, and protection status, and "or" inside the inner condition for altitude. Note that we needed to use brackets for defining the logical condition related to altitude.

In our example, we used the operator "or," and we got the result we wanted, but there is an alternative approach. Because there are three different altitude ranges ("0-500", "501-1000" and "1001-1500"), instead of using the form "0-500" or "501-1000," we could also define the condition not equal to "1001-1500". As we have already seen, the symbol for not equal to is "!=", so our command could also look like this:

```
#Subset function - more conditions
subset(df, size>10 & altitude!="1001-1500" & protection==T)

id          species  size    altitude  protection
1    1    Species 1   50       0-500       TRUE
2    2    Species 2   25       0-500       TRUE
3    3    Species 3   30     501-1000      TRUE
4    4    Species 4   45     501-1000      TRUE
```

The result we get is the same.

1.3.3   Removing Elements from a Data Frame

Removing elements from a data frame is similar to the procedure for selecting elements. Again (same as when removing elements from a vector), we use the symbol "-". Here are some examples; to remove the last row from the data frame named **df**, we do the following:

```
#Removing a row
df[-12,]
```

	id	species	size	altitude	protection
1	1	Species 1	50	0-500	TRUE
2	2	Species 2	25	0-500	TRUE
3	3	Species 3	30	501-1000	TRUE
4	4	Species 4	45	501-1000	TRUE
5	5	Species 5	2	501-1000	TRUE
6	6	Species 6	70	1001-1500	FALSE
7	7	Species 7	22	1001-1500	FALSE
8	8	Species 8	20	1001-1500	FALSE
9	9	Species 9	10	1001-1500	FALSE
10	10	Species 10	45	1001-1500	FALSE
11	11	Species 11	22	0-500	FALSE

The original data frame had 12 rows, so we specified this number in the square brackets, according to the rule. The same result could be obtained by applying the following command:

```
#Removing a row
df[-c(length(row(df)),]
```

This command is universal for removing the last row, because it indicates removing the row whose value is equal to the length of all rows.

We can also remove several rows at one time using one command. For example, to remove the first seven rows and the tenth row, we apply:

```
#Removing several rows
df[-c(1:7,10),]
```

	id	species	size	altitude	protection
8	8	Species 8	20	1001-1500	FALSE
9	9	Species 9	10	1001-1500	FALSE
11	11	Species 11	22	0-500	FALSE
12	12	Species 12	56	0-500	FALSE

The command to remove columns has a similar structure. To delete the first, third, and fourth column, we apply:

```
#Removing columns
df[,-c(1,3:4)]
```

	species	protection
1	Species 1	TRUE
2	Species 2	TRUE
3	Species 3	TRUE
4	Species 4	TRUE
5	Species 5	TRUE
6	Species 6	FALSE
7	Species 7	FALSE

```
8    Species 8     FALSE
9    Species 9     FALSE
10   Species 10    FALSE
11   Species 11    FALSE
12   Species 12    FALSE
```

Another option for removing columns from a data frame is based on using their name and the function "subset:"

```
#Removing columns by their name
subset(df, select=-c(altitude, protection))

     id    species  size
1    1   Species 1    50
2    2   Species 2    25
3    3   Species 3    30
4    4   Species 4    45
5    5   Species 5     2
6    6   Species 6    70
7    7   Species 7    22
8    8   Species 8    20
9    9   Species 9    10
10  10  Species 10    45
11  11  Species 11    22
12  12  Species 12    56
```

Last, we show how to remove certain columns and rows at the same time. For example, to remove rows 2 to 9, and columns 1 and 3, we will apply:

```
#Removing rows and columns
df[-(2:9), c(1,3)]

     id  size
1    1   50
10  10   45
11  11   22
12  12   56
```

1.3.4   Adding Elements to a Data Frame

In this section, we demonstrate how to add columns and rows to a data frame, and how to merge two data frames. All the examples use the data frame named **df** from the previous section. Let's say that we want to add a new column to the data frame **df**. First, we create a new element:

```
#Creating a new element - vector
year<-c(rep(2018:2019, 6))
year

[1] 2018 2019 2018 2019 2018 2019 2018 2019 2018 2019 2018 2019
```

The new column "year" might, for example, indicate when the field work has been performed. Now, we add this numerical vector to the data frame **df**, by using the function "cbind," and it will appear as a new column:

```
#Adding a column
cbind(df, year)
```

	id	species	size	altitude	protection	year
1	1	Species 1	50	0-500	TRUE	2018
2	2	Species 2	25	0-500	TRUE	2019
3	3	Species 3	30	501-1000	TRUE	2018
4	4	Species 4	45	501-1000	TRUE	2019
5	5	Species 5	2	501-1000	TRUE	2018
6	6	Species 6	70	1001-1500	FALSE	2019
7	7	Species 7	22	1001-1500	FALSE	2018
8	8	Species 8	20	1001-1500	FALSE	2019
9	9	Species 9	10	1001-1500	FALSE	2018
10	10	Species 10	45	1001-1500	FALSE	2019
11	11	Species 11	22	0-500	FALSE	2018
12	12	Species 12	56	0-500	FALSE	2019

```
#Alternatively
data.frame(df, year)
```

In addition to adding a column, we can also merge two data frames using the function merge. There are many practical ways to merge data frames, but generally the two data frames should share one or more columns that include a similar grouping variable. The grouping variable is usually a character or integer. If there are missing groups in one data frame the resulting merged data frame will by default exclude those groups, but this behavior can be changed with the "all" argument in the merge function. In our example, we create a new data frame with columns "id" and "category." The column "id" exists in both **df** and **df.category**, and the "category" is a new column that we intend to add to the data frame **df**.

```
#Creating a data frame
df.category<-data.frame(id=1:12, category=c("a", "b", "c"))
df.category
```

	id	category
1	1	a
2	2	b
3	3	c
4	4	a
5	5	b
6	6	c
7	7	a
8	8	b
9	9	c
10	10	a
11	11	b
12	12	c

Now we can merge the two data frames **df** and **df.category** with the function "merge":

```
#Merging two data frames
merge(df, df.category, by="id")
```

	id	species	size	altitude	protection	category
1	1	Species 1	50	0-500	TRUE	a
2	2	Species 2	25	0-500	TRUE	b
3	3	Species 3	30	501-1000	TRUE	c
4	4	Species 4	45	501-1000	TRUE	a
5	5	Species 5	2	501-1000	TRUE	b
6	6	Species 6	70	1001-1500	FALSE	c
7	7	Species 7	22	1001-1500	FALSE	a
8	8	Species 8	20	1001-1500	FALSE	b
9	9	Species 9	10	1001-1500	FALSE	c
10	10	Species 10	45	1001-1500	FALSE	a
11	11	Species 11	22	0-500	FALSE	b
12	12	Species 12	56	0-500	FALSE	c

Next, we show how to add a row to the data frame **df**. First, we need to create a row to be added. Let's say that we want to create a new data frame that consists of the minimum values of each column in the data frame **df**:

```
#Creating a new data frame
df.row<-apply(df,2,min)
df.row
```

id	species	size	altitude	protection
"1"	"Species 1"	" 2"	"0-500"	" TRUE"

We can add this data frame to the date frame **df** by using the function "rbind":

```
#Adding a new row
rbind(df, df.row)
```

	id	species	size	altitude	protection
1	1	Species 1	50	0-500	TRUE
2	2	Species 2	25	0-500	TRUE
3	3	Species 3	30	501-1000	TRUE
4	4	Species 4	45	501-1000	TRUE
5	5	Species 5	2	501-1000	TRUE
6	6	Species 6	70	1001-1500	FALSE
7	7	Species 7	22	1001-1500	FALSE
8	8	Species 8	20	1001-1500	FALSE
9	9	Species 9	10	1001-1500	FALSE
10	10	Species 10	45	1001-1500	FALSE
11	11	Species 11	22	0-500	FALSE
12	12	Species 12	56	0-500	FALSE
13	1	Species 1	2	0-500	TRUE

Adding a row to a data frame is also possible if we create new rows as lists. Lists are one- or two-dimensional arrays that can contain elements of different types. To create a list, we use the command "list(elements)." Here, we create a list that has the same structure as the data frame **df**:

```
#Creating a list
row.list<-list(15, "Species 8", 28, "501-1000", T)
row.list
```

We can add the list we just created to the data frame **df**. In this example, we show how to add a list to a subset of the data frame **df** that includes only the first three rows. The procedure is similar to the one previously described; we use the function "rbind," and the command for extracting elements [nrows,]:

```
#Adding a list to the data frame (selecting rows)
rbind(df[1:3,],row.list)

   id  species  size  altitude  protection
1   1  Species 1   50    0-500      TRUE
2   2  Species 2   25    0-500      TRUE
3   3  Species 3   30  501-1000     TRUE
4  15  Species 8   28  501-1000     TRUE
```

1.3.5 Additional Operations on Data Frames

In this section, we show how to apply several more functions for manipulating data frames. One function that is commonly used is the function "cut." This function enables grouping the data into different categories. In the next example, we apply the "cut" function to the column named "size." When applying this function, we can define the number of groups we want to have, or we can define the precise thresholds for each group.

First, we apply the "cut" function to define the number of groups. For example, based on the size of populations, we want to define three groups labeled as "small", "medium" and "large:"

```
#Cut function
size.group<-cut(df$size, breaks = 3,
labels=c("small", "medium", "large"))

size.group
```

Now, we add this vector to the data frame **df** by applying the function "cbind:"

```
#Adding columns
cbind(df, size.group)

   id  species  size  altitude  protection  size.group
1   1  Species 1   50    0-500      TRUE        large
2   2  Species 2   25    0-500      TRUE       medium
3   3  Species 3   30  501-1000     TRUE       medium
4   4  Species 4   45  501-1000     TRUE       medium
```

5	5	Species 5	2	501-1000	TRUE	small
6	6	Species 6	70	1001-1500	FALSE	large
7	7	Species 7	22	1001-1500	FALSE	small
8	8	Species 8	20	1001-1500	FALSE	small
9	9	Species 9	10	1001-1500	FALSE	small
10	10	Species 10	45	1001-1500	FALSE	medium
11	11	Species 11	22	0-500	FALSE	small
12	12	Species 12	56	0-500	FALSE	large

In the above example, the command itself found thresholds for each group, and all we had to define was the number of groups we wanted to get. However, there is also an option to define a specific threshold. For example, we might say that population size <= 30 is considered "regular," and size > 30 is considered "remarkable." We want two groups, which we can define in the following way:

```
#Cut function
size.group.t<-cut(df$size, breaks=c(0,30,100),
labels=c("regular", "remarkable"))

size.group.t
```

The syntax "breaks=c(0,30,100)" means that we created two groups: the first one takes values in the range (0,30] and the second one belongs to the range (30, 100]. Note that we had to define an upper limit (100), and to pay attention to the choices of round and square brackets needed to define the groups.

We can add this new vector to the data frame, again applying the function "cbind:"

```
#Adding columns
cbind(df, size.group.t)
```

	id	species	size	altitude	protection	size.group.t
1	1	Species 1	50	0-500	TRUE	remarkable
2	2	Species 2	25	0-500	TRUE	regular
3	3	Species 3	30	501-1000	TRUE	regular
4	4	Species 4	45	501-1000	TRUE	remarkable
5	5	Species 5	2	501-1000	TRUE	regular
6	6	Species 6	70	1001-1500	FALSE	remarkable
7	7	Species 7	22	1001-1500	FALSE	regular
8	8	Species 8	20	1001-1500	FALSE	regular
9	9	Species 9	10	1001-1500	FALSE	regular
10	10	Species 10	45	1001-1500	FALSE	remarkable
11	11	Species 11	22	0-500	FALSE	regular
12	12	Species 12	56	0-500	FALSE	remarkable

Now, we show how to perform different operations on a data frame, taking into account certain conditions. For example, we can find a maximum population size for species with status protection equal to true.

```
#Manipulating data frame
protected<-subset(df, protection==T)
max(protected$size)
```

[1] 50

Another example is to count the number of species with protection status equal to true:

```
#Manipulating data frame
length(which(df$protection==T))
```

```
[1] 5
```

We can extend that condition, and count the number of species that are protected and registered at the altitude range "0-500:"

```
#Manipulating data frame
length(which(df$protection==T & df$altitude=="0-500"))
```

```
[1] 2
```

Another example is finding a maximum size of population for each altitude range. For that purpose, we will use the function "aggregate." The general syntax for this function is aggregate(data, by=list(condition), function) and in our example it will look like:

```
#Manipulating data frame
df3<-aggregate(df$size, by=list(altitude), FUN=max)
df3
```

```
      Group.1    x
1       0-500   56
2   1001-1500   70
3    501-1000   45
```

To add column names to data frame df3, and then reorder it by maximum size of population:

```
#Assigning names to the data frame
names(df3)<-c("altitude", "max population size")
df3

#Reordering the data frame
df3[order(df3$`max population size`),]
```

```
      altitude   max population size
3    501-1000                     45
1       0-500                     56
2   1001-1500                     70
```

A similar example is to calculate average size of populations by the protection status:

```
#Manipulating data frame
df4<-aggregate(df$size, by=list(protection), FUN=mean)
df4
```

```
   Group.1     x
1    FALSE   35.0
2     TRUE   30.4
```

Again, to modify the data frame by assigning column names:

```
#Manipulating data frame
names(df4)<-c("protection", "average population size")
df4
```

	protection	average population size
1	FALSE	35.0
2	TRUE	30.4

1.4   Missing Data

One important question is, how to deal with missing data within a data frame? For demonstration, we again use our data frame **df**. However, in the original data frame, there were no missing elements, so we need to make some changes to **df** first.

Turning a complete data frame into an incomplete one can happen in real studies when processing data, especially in cases in which one gets a result that does not make sense (for example, getting negative values for some measurements). Therefore, instead of keeping values that are wrong or illogical, we can assign them the value of "NA."

In the example of the data frame **df**, we can replace size of population that is equal to 45 with the value "NA." We will also create a new data frame named **df.na** to use later. Here is how we do it:

```
#Replacing values in a data frame
df$size[df$size==45]<-NA
df.na<-df
df.na
```

	id	species	size	altitude	protection
1	1	Species 1	50	0-500	TRUE
2	2	Species 2	25	0-500	TRUE
3	3	Species 3	30	501-1000	TRUE
4	4	Species 4	NA	501-1000	TRUE
5	5	Species 5	2	501-1000	TRUE
6	6	Species 6	70	1001-1500	FALSE
7	7	Species 7	22	1001-1500	FALSE
8	8	Species 8	20	1001-1500	FALSE
9	9	Species 9	10	1001-1500	FALSE
10	10	Species 10	NA	1001-1500	FALSE
11	11	Species 11	22	0-500	FALSE
12	12	Species 12	56	0-500	FALSE

1.4.1   Removing Missing Values

Because we have elements missing in the column "size" in the data frame **df**, we cannot perform some standard operations. Let's try, for example, to calculate the sum of the column "size:"

```
#Summing a column with missing values
sum(df$size)
```

[1] NA

Therefore, we need to remove missing values, and then calculate the sum. We can perform this with a single command:

```
#Summing a column without missing values
sum(df$size, na.rm=T)
```

[1] 307

The function we used, "na.rm=T," enabled removing missing values, so we could get the final result.

To list all rows with missing elements, we apply the following command using the syntax "!complete.cases(data)":

```
#Listing rows with missing elements
df[!complete.cases(df),]
```

	id	species	size	altitude	protection
4	4	Species 4	NA	501-1000	TRUE
10	10	Species 10	NA	1001-1500	FALSE

This command is especially useful when dealing with large datasets with missing elements.

In our first example of operating with missing data, we removed missing values from a column by applying "na.rm=T," but we can also remove the entire row which contains missing element(s), and that is done by the command "na.omit(data)":

```
#Removing rows with missing elements
na.omit(df)
```

	id	species	size	altitude	Protection
1	1	Species 1	50	0-500	TRUE
2	2	Species 2	25	0-500	TRUE
3	3	Species 3	30	501-1000	TRUE
5	5	Species 5	2	501-1000	TRUE
6	6	Species 6	70	1001-1500	FALSE
7	7	Species 7	22	1001-1500	FALSE
8	8	Species 8	20	1001-1500	FALSE
9	9	Species 9	10	1001-1500	FALSE
11	11	Species 11	22	0-500	FALSE
12	12	Species 12	56	0-500	FALSE

Both functions, "na.rm=T" and "na.omit(data)," are applied when processing the results, and it is up to the researcher as to which one they use.

1.4.2 Replacing Missing Values

Sometimes, instead of removing missing values, we might want to replace them. Some typical examples are to calculate the average, median, or mode value for the complete data, and use one of these values to replace missing values. Let's have a closer look at this issue. First, we calculate the mode for complete values in the column "size".

```
#Creating a vector with complete cases
size.complete<-df$size[complete.cases(df$size)]
size.complete

#Finding a mode value for complete cases
size.complete<-factor(size.complete)
mode<-rev(sort(summary(size.complete)))
mode<-mode[1]
mode

22
 2
```

As a result, we see that the modal value is 22, and this value repeats two times within the column "size." We can replace missing values with the modal value as follows:

```
#Filling missing elements with the mode value
df$size[!complete.cases(df$size)]<-22
df$size

[1] 50 25 30 22  2 70 22 20 10 22 22 56
```

As another example, we can replace missing values with the average value of complete cases in the column.

```
#Finding the average value of complete cases
mean<-mean(df$size, na.rm=T)
mean

[1] 30.7

#Filling missing elements with the average value
df$size[!complete.cases(df$size)]<-mean
df$size

[1] 50.0 25.0 30.0 30.7  2.0 70.0 22.0 20.0 10.0 30.7 22.0 56.0
```

A similar procedure works for replacing missing values with the median value of complete cases:

```
#Finding the median value of complete cases
median<-median(df$size, na.rm=T)
median

[1] 23.5

#Filling missing elements with the median value
df$size[!complete.cases(df$size)]<-median
df$size

[1] 50.0 25.0 30.0 23.5  2.0 70.0 22.0 20.0 10.0 23.5 22.0 56.0
```

If we now display the data frame **df**, in which missing values have been replaced by median values, we get:

```
#Data frame with filled elements (median values)
df

     id    species    size    altitude    protection
1    1     Species 1   50.0      0-500        TRUE
2    2     Species 2   25.0      0-500        TRUE
3    3     Species 3   30.0    501-1000       TRUE
4    4     Species 4   23.5    501-1000       TRUE
5    5     Species 5    2.0    501-1000       TRUE
6    6     Species 6   70.0   1001-1500      FALSE
7    7     Species 7   22.0   1001-1500      FALSE
8    8     Species 8   20.0   1001-1500      FALSE
9    9     Species 9   10.0   1001-1500      FALSE
10   10    Species 10  23.5   1001-1500      FALSE
11   11    Species 11  22.0      0-500       FALSE
12   12    Species 12  56.0      0-500       FALSE
```

Here is one last point about missing values. There is a conventional threshold regarding the number of missing values in a row or column that should be taken into account. If we have large datasets, a threshold of 5% missing values is considered "safe." For cases in which there are more than 5% missing values, though, it is recommended that we discard that column or row from the dataset.

To check the percentage of missing elements in columns in the data frame **df.na,** we apply the following function:

```
#Checking percentage of missing elements in a column
per.missing<-function(x){sum(is.na(x))/length(x)*100}
apply(df.na, 2, per.missing)

id          species         size    altitude    protection
0.00000     0.00000     16.66667    0.00000      0.00000
```

Here, we see that 16.67% of elements are missing in the column "size." Because this is a small data set, we don't necessarily want to apply the 5% rule, but in cases of large data sets with this percentage of missing elements, the recommendation would be to drop that column.

A similar procedure can be applied to check the percentage of missing elements within rows. Here, we just modify the element of the function "apply:"

```
#Checking percentage of missing elements in a row
miss.rows<-apply(df.na, 1, per.missing)
cbind(id, miss.rows)

        id    miss.rows
[1,]    1         0
[2,]    2         0
[3,]    3         0
[4,]    4        20
[5,]    5         0
[6,]    6         0
[7,]    7         0
```

```
     [8,]    8              0
     [9,]    9              0
    [10,]   10             20
    [11,]   11              0
    [12,]   12              0
```

We can see that for species with the "id 4" and "id 10," 20% of elements have missing data, and the rest of the row data is complete.

1.5   Outliers

When processing results, we often get outlier values (extreme data values that do not appear to fit the distribution of most other values), and the question is, what to do with them if they occur? There are basically two options; the first is to transform the vector, if we believe that one or more elements are extreme but valid, and the second is to remove the outliers. Let's analyze an example with outliers; let's say that we have measured the height of mature poplar trees:

```
#Measured height of poplar trees [meters]
tree.height.meters<-c(27,30,32,28,35,42,27,150)
```

Right away, we can tell that the value of 150 is an outlier, but we can also see this more clearly by creating an appropriate box plot:

```
#Creating a box plot
boxplot(tree.height.meters, ylab = "height [m]")
```

Figure 1.1 shows the results.

1.5.1   Transforming Outliers

When dealing with outliers that we think may very well be valid, the most general procedure is to transform the vector values into logarithmic values. In our example, that would look like:

Fig. 1.1 Height of trees [m], boxplot

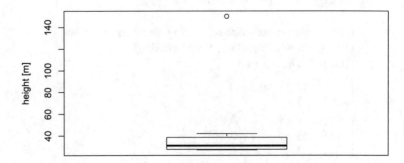

```
#Transforming values using log funtion
log.tree.height<-log(tree.height.meters)
log.tree.height<-round(log.tree.height,2)
log.tree.height
```

[1] 3.30 3.40 3.47 3.33 3.56 3.74 3.30 5.01

Now, we can plot the transformed values:

```
#Plotting transformed values
plot(log.tree.height, xlab="measurement")
```

And we get (Fig. 1.2):

From this figure, we see that the outlier effect is much less pronounced with the function "log," but we can also analyze some descriptive statistics (average, median and standard deviation) for the initial and transformed values.

First, we create a matrix with the initial and transformed data:

```
#Creating a matrix (initial and transformed values)
it<-matrix(c(tree.height.meters,log.tree.height), ncol=2, byrow=F)
colnames(it)<-c("initial", "transformed")
it
```

And then define multiple functions to apply over the matrix **it**:

```
#Defining multiple functions
multi.fun <- function(x) {c(mean = mean(x),
median = median(x), sd = sd(x))}
```

In the next step, we apply these functions to each column of the matrix **it**:

```
#Apply multiple functions
apply(it,2, multi.fun)
```

Fig. 1.2 Height of trees, logarithmic transformation

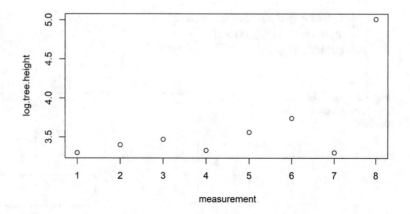

And we will get the result:

	initial	transformed
mean	46.3750	3.636808
median	31.0000	3.433467
sd	42.1729	0.574990

These results demonstrate that using the "log" function helps with outliers, but we next show how to exclude them from the database.

1.5.2   Removing Outliers

To remove outliers from the vector **tree.height.meters,** first we rename this vector to **initial**, just to have the commands easier to read. Now, the shortest way to remove outliers is:

```
#Removing outliers
initial<-tree.height.meters

rm.out<-initial[!initial%in%boxplot(initial)$out]
rm.out
```

And we get the result:

[1] 27 30 32 28 35 42 27

Now, we create a new boxplot for these last results:

```
#Creating a box plot
boxplot(rm.out, ylab="height [m]")
```

We get Fig. 1.3:

We can also calculate descriptive statistics for the vector **rm.out**. We use the package "pastecs," but first we need to install it:

```
#Installing the package "pastecs"
install.packages("pastecs")
library(pastecs)
```

Fig. 1.3 Height of trees [m], after removing outliers

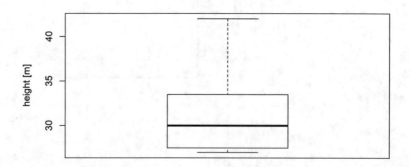

Now, we apply one function from this package, called "stat.desc":

```
#Calculating parameters of descriptive statistics
stat.desc(rm.out)
round(stat.desc(rm.out),2)
```

And we get this output:

nbr.val	nbr.null	nbr.na	min	max	range	sum
7.00	0.00	0.00	27.00	42.00	15.00	221.00
median	mean	SE.mean	CI.mean.0.95	var	std.dev	coef.var
30.00	31.57	2.06	5.03	29.62	5.44	0.17

We recommend using the "log" function, when we have extreme values that we consider to be valid. If extreme values are likely to be a mistake in measuring or processing data, it is best to remove those values, and deal with the remaining elements in the data set.

Numerical Analysis

2

In this chapter we deal with basic and more advanced numerical analysis and with graphical representation of the results. As the input data, we create vectors, matrices, and data frames directly in R (using the procedures described in Chap. 1), but we also import data files already created in Excel. R supports creating input data within the program itself, but also supports importing data files from Excel, SPSS, SAS, etc., and we demonstrate both options.

The first section in this chapter presents procedures for basic operations without using R packages, and the subsequent section shows procedures for performing calculations with appropriate R packages. Examples shown in this chapter are related to determining the biological spectrum of flora, calculating alpha and beta diversity indices, calculating the basic statistical parameters, etc. This chapter explains how to present the results graphically, and we create pie charts, group and stacked bar charts, dendrograms, gradient matrices, etc.

2.1 Basic Numerical Analyses

In this section, we demonstrate how to perform some basic ecological analyses. The first example determines the biological spectrum of species within an area. Along with explaining the calculations, we demonstrate how to present the results graphically. The second example determines the share of species within a community for two different time periods, and once again the results are supported by different graphical outputs. In the first example, we import an Excel file with input data, and in the second example we create a data frame in R.

2.1.1 Determining Biological Spectrum

As our first example, we calculate the biological spectrum of flora in a region. As a reminder, the biological spectrum of flora describes the share of different life forms of plants (phanerophytes, chamaephytes, hemicryptophytes, geophytes and terophytes) within a certain area (Raunkiaer 1934).

© Springer Nature Switzerland AG 2020
M. Lakicevic et al., *Introduction to R for Terrestrial Ecology*,
https://doi.org/10.1007/978-3-030-27603-4_2

For this example, we use the Excel file named Biological.spectrum.xlxs.[1] After launching R, we import the data file for this Chapter, by clicking on "Import Dataset/From Excel," then choose the file path – folder location on the reader's machine, and then click "Import," (a preview of the file is provided in Appendix 1). After that, we use the R command "attach" with the name of the file and the import is done:

```
#Attaching an Excel file
attach(Biological_spectrum)
bs<-Biological_spectrum
bs
```

We can explore our data first. To check the dimensions of our data, we use the command "dim():"

```
#Checking dimensions of a data frame
dim(bs)
```

We get the result:

[1] 63 2

This means that our table has 63 rows and 2 columns.

To inspect our data frame in more detail, we use the function "str():"

```
#Checking structure of the data frame
str(bs)
```

And we get this result:

```
Classes 'tbl_df', 'tbl' and 'data.frame':   63 obs. of  2 variables:
 $ Species  : chr  "Agrostis canina L." "Anemone nemorosa L." "Antennaria dioica (L.) Gaertn."
"Anthoxanthum odoratum L." ...
 $ Life.form: chr  "H" "G" "H" "H" ...
```

These results indicate two columns named "Species" and "Life.form" with several elements of these columns. To list the first six rows of our data frame, we use the command "head()":

```
#Listing the first six rows
head(bs)
```

And we get the output:

Species	Life.form
<chr>	<chr>
1 Agrostis canina L.	H
2 Anemone nemorosa L.	G

---

[1]See the introductory section, About this Book, for information on how to download Excel files mentioned here and subsequently.

3 Antennaria dioica (L.) Gaertn.   H
4 Anthoxanthum odoratum L.   H
5 Arrhenatherium elatius (L.) J. &. C. Presl.   H
6 Asperula cynanchica L.   H

Similarly, to list the last six rows from the data frame **bs**, we use the function "tail()":

```
#Listing the last six rows
tail(bs)
```

We get the list below:

Species	Life.form
<chr>	<chr>
1 Thymus pulegioides L.	Ch
2 Trifolium alpestre L.	H
3 Trifolium pratense L.	H
4 Vaccinium myrtillus L.	Ch
5 Veronica chamaedrys L.	Ch
6 Viola sylvestris Lam.	H

Using the function "attach" shortens the process of selecting columns. For example, to select the column "Life.form," instead of using bs$Life.form, we can simply type Life.form. In these terms, if we want to list unique elements in this column, we apply:

```
#Listing unique elements from a column
unique(Life.form)
```

And we get:

[1] "H" "G" "P" "T" "Ch"

These labels represent the life forms chamaephytes, geophytes, phanerophytes, terophytes and hemicryptophytes, respectively.

To create a data frame containing all terophytes, we apply the function "subset:"

```
#Create a new data frame with the terophytes
terophytes<-subset(bs, Life.form=="T")
terophytes
```

And we get a new data frame:

A tibble: 5 x 2

Species	Life.form
<chr>	<chr>
1 Campanula patula L.	T
2 Campanula sparsa Friv.	T
3 Rosa tomentosa Sm.	T
4 Senecio viscosus L.	T
5 Solanum hispidum Pers.	T

From the last command, we see that the data frame **bs** has five species belonging to the category tero-phytes, and we can check their Latin names. We can repeat the same procedure with other life forms.

In previous steps, we got acquainted with the input file, and now we can proceed with calculating the biological spectrum of flora. To do this, we shall count the number of species belonging to each category of life form, and then convert these values into percentages. We can count the number of species in each category of life form if we turn the data in the column "Life.form" into a factor, and then check the summary:

```
#Turning Life.form into a factor
Life.form<-factor(Life.form)

#Summary function
summary(Life.form)
```

This way, we get:

```
Ch G H  P  T
 8  3 34 13  5
```

To turn these values into their respective share (percentages), we apply "prop.table()," and in that way we get the biological spectrum of flora for our data set:

```
#Calculating biological spectrum
bs.results<-prop.table(summary(Life.form))*100
bs.results<-round(bs.results,2)
bs.results
```

The results we get are:

```
Ch     G     H     P     T
12.70  4.76 53.97 20.63  7.94
```

This last result means that, in our example, we have 12.7% chamaephytes, 4.76% geophytes, and so on. We can also sort these values in descending order:

```
#Putting the results into a descending order
bs.results[rev(order(bs.results))]
```

And we get:

```
   H     P    Ch     T     G
53.97 20.63 12.70  7.94  4.76
```

Now, it is easier to see that hemicryptophytes prevail with 53.97%, and the geophytes are the smallest group with 4.76%.

We can present these results graphically using the function "pie." First, we transform **bs.results** into a matrix, and then set up the labels to be displayed on the pie chart:

```
#Transforming the results into a matrix
bs.results<-as.matrix(bs.results)

#Setting up the labels
labels<-paste(c(rownames(bs.results)), ":",
bs.results[,1])
```

```
labels<-paste(labels, "%", sep = "")
#Adjusting color and plotting the results
pie(bs.results, labels = labels, col=topo.colors(5))
```

We get the following result (Fig. 2.1), representing the biological spectrum of flora from our data set.

Fig. 2.1 Biological spectrum of flora

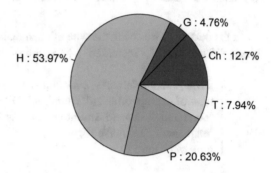

2.1.2 Determining Share of Species Within a Community

In the next example, we analyze the following problem. Let's say that we have determined the number of species and the number of individuals belonging to each species within a small plot. Furthermore, let's say that we performed the field work two times, in 2014 and 2019, and that we want to compare the results. Table 2.1 presents our data.

We will create the corresponding data frame in the following way:

```
#Creating a data frame with input data
df<-data.frame(species=paste("Species",LETTERS[1:7]),
               number.2014=c(25,40,35,8,62,33,12),
               number.2019=c(35,45,35,15,42,26,5))
df
```

First, we analyze the values in the data frame **df**. To start, we check to see if the number of individuals for each species has changed. For that purpose, we use the function "ifelse," and this is how this function looks when we have two outputs: "changed" and "remained equal" (note that "!=" stands for unequal):

Table 2.1 Species and the size of their populations in two time slots

Species	Number of individuals in 2014	Number of individuals in 2019
Species A	25	35
Species B	40	45
Species C	35	35
Species D	8	15
Species E	62	42
Species F	33	26
Species G	12	5

```
#The function "ifelse" - checking changes in number of individuals
change<-ifelse(df$number.2014!=df$number.2019,
"changed", "equal")
change
```

We get the following output:

```
[1] "changed" "changed" "equal"   "changed" "changed" "changed" "changed"
```

In the next step, we check the type of change. Namely, we want to test three possible outputs ("no change", "decreased", "increased"). Now, the syntax is a bit more complex:

```
#The function "elseif" - checking type of changes in number of individuals
type.of.change<-ifelse(df$number.2014==df$number.2019,
"no change", ifelse(df$number.2014<df$number.2019,
"increased", "decreased"))
type.of.change
```

We can add these results to the data frame **df**:

```
#Adding a column (type of change)
df.change<-cbind(df, type.of.change)
df.change
```

Now we have the result:

```
  species number.2014 number.2019 type.of.change
1 Species A          25          35      increased
2 Species B          40          45      increased
3 Species C          35          35      no change
4 Species D           8          15      increased
5 Species E          62          42      decreased
6 Species F          33          26      decreased
7 Species G          12           5      decreased
```

As the final step for analyzing changes in the number of individuals for each species, we apply the "count" function from the "dplyr" package:

```
#Installing the package "dplyr"
install.packages("dplyr")
library(dplyr)

#Counting type of changes
df.change%>%count(type.of.change)
```

The result is:

```
type.of.change    n
  <fct>         <int>
1 decreased       3
2 increased       3
3 no change       1
```

To graphically present the number of species in 2014 and 2019, we create two different plots: stacked and grouped bar. When defining values to be placed on the x and y axes, we need to have unique values, so we cannot specify two different columns to be printed on the y axis. Therefore, we need to make some adjustments to the data frame **df** before printing the results. In fact, we need to create a new data frame with one column showing the number of individuals in which the first half of rows show the number of individuals in 2014 and then the second half the number in 2019. This transformation is easily done with the function "melt" from the package "reshape":

```
#Installing the package "reshape"
install.packages("reshape")
library(reshape)

#Re-organizing data frame – function "melt"
df.melted<-melt(df, id.vars ="species")
df.melted
```

Here is the output we get:

```
   species        variable   value
1  Species A   number.2014     25
2  Species B   number.2014     40
3  Species C   number.2014     35
4  Species D   number.2014      8
5  Species E   number.2014     62
6  Species F   number.2014     33
7  Species G   number.2014     12
8  Species A   number.2019     35
9  Species B   number.2019     45
10 Species C   number.2019     35
11 Species D   number.2019     15
12 Species E   number.2019     42
13 Species F   number.2019     26
14 Species G   number.2019      5
```

We have created the data frame named **df.melted**, but we can make some additional changes to it. We will rename columns and elements of the column "Year:"

```
#Renaming columns and columns' elements
names(df.melted)<-c("Species", "Year", "Number")
df.melted$Year<-as.character(df.melted$Year)
df.melted$Year[df.melted$Year=="number.2014"]<-2014
df.melted$Year[df.melted$Year=="number.2019"]<-2019
df.melted
```

Now, we have everything set for plotting the results. We use the "geom_bar" function from the package "ggplot2," and create a stacked bar. The procedure is similar to the one for creating a "regular" bar plot, but we use the additional command "position = position_stack()":

```
#Installing the package "ggplot2"
install.packages("ggplot2")
library(ggplot2)
```

```
#Ploting the results - stacked bar (by year)
library(ggplot2)
p<-ggplot(df.melted, aes(x=Year, y=Number, fill=Species))+
geom_bar(stat="identity")

#Adding values
p<-p+ geom_text(aes(label=Number), size= 4,
position=position_stack(vjust=0.5))

#Adjusting color
p+scale_fill_brewer(palette = "Set3")
```

Figure 2.2 shows the results.

We can present the same results using the grouped bar. In that case, we use the following commands:

```
#Ploting the results - grouped bar
p1<-ggplot(df.melted, aes(x=factor(Species), y=Number, fill=Year)) + geom_
bar(stat="identity", position = "dodge")

#Adding values
p1<-p1+geom_text(aes(label=Number),
position=position_dodge(width=0.9), vjust=-0.5)

#Changing colors
p1<-p1+scale_fill_brewer(palette = "Set1")

#Assigning names - x and y axis
p1+xlab("")+ylab("Number")
```

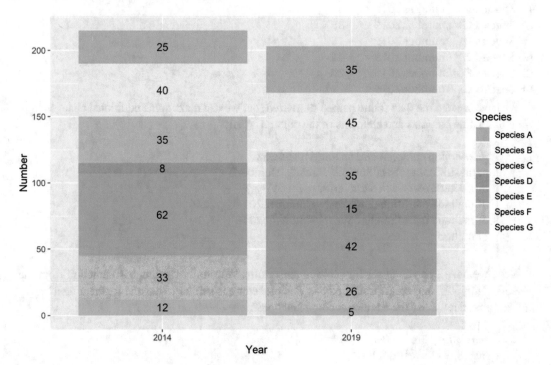

Fig. 2.2 Number of species' individuals—stacked bar format

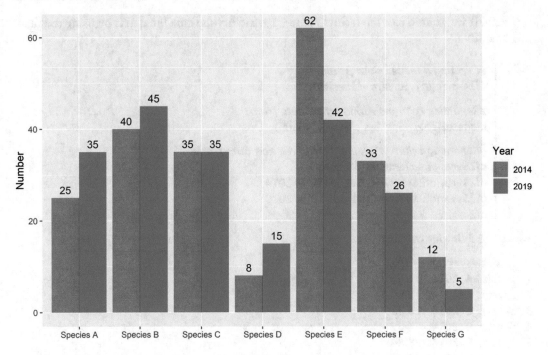

Fig. 2.3 Number of species' individuals—grouped bar format

And we have a new representation (Fig. 2.3):

Next, we calculate the share of species in both 2014 and 2019, and create a new data frame containing these values. The easiest way to calculate the share of species is by using the function "prop.table():"

```
#Calculating share of species (2014 and 2019)
share.2014<-round(prop.table(df$number.2014)*100,2)
share.2019<-round(prop.table(df$number.2019)*100,2)

#Creating a data frame with output data
df1<-data.frame(Species=df$species, share.2014, share.2019)
df1
```

And we get the following results:

```
  Species  share.2014  share.2019
1 Species A     11.63       17.24
2 Species B     18.60       22.17
3 Species C     16.28       17.24
4 Species D      3.72        7.39
5 Species E     28.84       20.69
6 Species F     15.35       12.81
7 Species G      5.58        2.46
```

Again, if we want to plot the results, we need to modify the data fame **df1**, by using the function "melt:"

```
#Creating a melted data frame
df2<-melt(df1, id.vars ="Species")

#Renaming columns within a new data frame
names(df2)<-c("Species", "year", "share")

#Renaming columns' elements within a new data frame
df2$year<-as.character(df2$year)
df2$year[df2$year=="share.2014"]<-2014
df2$year[df2$year=="share.2019"]<-2019
df2
```

We get the following output:

```
    Species   year   share
1   Species A  2014   11.63
2   Species B  2014   18.60
3   Species C  2014   16.28
4   Species D  2014    3.72
5   Species E  2014   28.84
6   Species F  2014   15.35
7   Species G  2014    5.58
8   Species A  2019   17.24
9   Species B  2019   22.17
10  Species C  2019   17.24
11  Species D  2019    7.39
12  Species E  2019   20.69
13  Species F  2019   12.81
14  Species G  2019    2.46
```

Now, we proceed with plotting the results. We use the "geom_bar" function and the "facet_wrap." The function "facet_wrap" allows printing the results by dividing them into separate panels, where each panel corresponds to a selected (categorical) variable. In this example, we show how to print the results based on the year and based on the species, by applying this function. First, we divide the results based on the year:

```
#Ploting the results - by year
p<-ggplot(data=df2, aes(x=Species, y=share, fill=Species))+
geom_bar(stat="identity")+ facet_wrap(~year)

#Adding values
p<-p+geom_text(aes(label=share), vjust=-0.2, size=3.5)

#Adjusting x and y axis (labels, ticks,...)
p+theme(axis.text.x=element_blank(),
        axis.ticks.x=element_blank()) + ylab("Share [%]")
```

We now have the output:

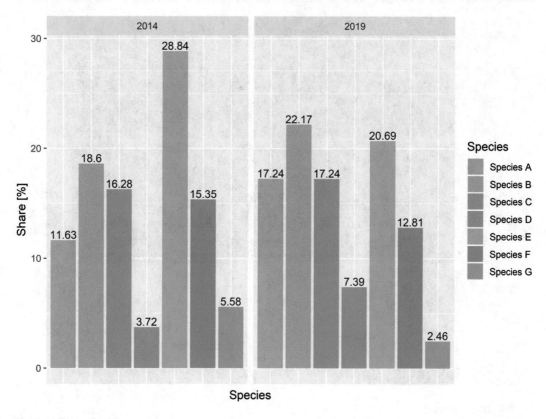

Fig. 2.4   Share of species

Figure 2.4 can be re-ordered, if we want to see the changes between 2014 and 2019, for each species separately:

```
#Ploting the results – by species
p1<-ggplot(data=df2, aes(x=year, y=share, fill=year))+
geom_bar(stat="identity")+
facet_wrap(~Species)

#Adding values
p1<-p1+geom_text(aes(label=share),
vjust=0.9, size=3.2)

#Adjusting names and labels on the x and y axis
p1+theme(axis.text.x=element_blank(),
         axis.ticks.x=element_blank())+
         ylab("Share [%]")+xlab("")
```

Now, we have this output (Fig. 2.5):

In this example, we have shown some basic operations, and some of the main types of graphical representation.

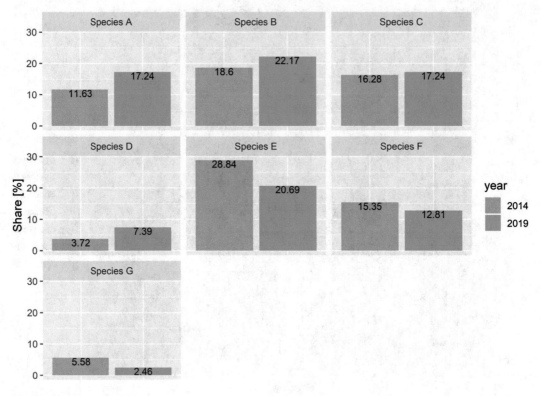

Fig. 2.5 Share of species

2.2   Advanced Numerical Analyses

In this section, we demonstrate how to calculate biodiversity indices with: (a) formulas, and (b) the "vegan" package. In addition, we present an example, and provide more detailed insight into numerical analysis in ecological tasks.

2.2.1   Biodiversity Indices

Biodiversity indices are divided into two main groups of alpha and beta indices. Alpha indices quantify biodiversity within a single area, whereas beta indices compare similarities related to biodiversity among different areas (Begon et al. 1996). Moreover, beta indices can be applied to a single area, but different time periods, and in that case these values measure changes that have occurred over time.

2.2.1.1 Calculating Alpha Biodiversity Indices with Formulas
In this section, we demonstrate how to calculate two alpha biodiversity indices: the Shannon and Simpson indices. We use the same data frame for both:

```
#Creating the input data frame
df.bd<-data.frame(Species=paste("Species", LETTERS[1:5]),
                  Number=c(25,40,50,15,22))
df.bd
```

Our data frame with input data is

```
    Species Number
1 Species A      25
2 Species B      40
3 Species C      50
4 Species D      15
5 Species E      22
```

Now, we can calculate both the Shannon and Simpson indices.

2.2.1.1.1 Shannon Index

The formula for calculating the value of the Shannon index (H') is:

$$H' = -\sum_{i=1}^{s} p_i \ln(p_i)$$

in which: p_i is the proportion of species i in the community, and s is the total number of species in the community (Magurran 2004).

Let's apply this formula to the data frame **df.bd**:

```
##Shannon index
#Calculating p
p<-round(prop.table(df.bd$Number),3)

#Adding "p" column to the original data frame
df.bd1<-cbind(df.bd, p)

#Calculating ln(p)
ln.p<-round(log(x=p),3)

#Adding "ln(p)" column to the data frame
df.bd1<-cbind(df.bd1, ln.p)

#Multiplying p and ln(p) and adding the column
df.bd1<-cbind(df.bd1, multipl=round(p*ln.p,3))
df.bd1
```

Here is the updated data frame:

```
    Species Number     p    ln.p   multipl
1 Species A     25  0.164  -1.808   -0.297
2 Species B     40  0.263  -1.336   -0.351
3 Species C     50  0.329  -1.112   -0.366
4 Species D     15  0.099  -2.313   -0.229
5 Species E     22  0.145  -1.931   -0.280
```

To calculate the Shannon index, we just need to sum the values in the column "multipl" and multiply it by -1.

```
#Calculating Shannon index
shannon<-sum(df.bd1$multipl)*(-1)
shannon
```

The result is:

[1] 1.523

Usually, values of the Shannon index fall in the range between 1.5 and 3.5 (Magurran 2004). Values of the Shannon index above 3.5, and in particular above 4, suggest that the community has high biodiversity. In our example, we had a value of 1.523, which suggests that this community cannot be considered as highly valuable in terms of biodiversity (Fedor and Zvaríková 2019).

2.2.1.1.2 Simpson Index

In the next example, we calculate the value of the Simpson index with the formula:

$$D = 1 - \sum_{i=1}^{S} p_i^2$$

in which: p_i is the proportion of species i in the community and s is a total number of species in the community (Magurran 2004).

Here are the steps for calculating the Simpson index for the data in the data frame **df.bd**:

```
##Simpson index
#Calculating p
p<-round(prop.table(df.bd$Number),3)

#Calculating p*p
p.2<-round(p*p,3)

#Calculating Simpson index
simpson<-1-sum(p.2)
simpson
```

The resulting value of the Simpson index is:

[1] 0.765

The value of the Simpson index is in the range [0, 1], and larger values correspond to more even distribution of species within a community.

Although biodiversity indices can be calculated without a package, as we have shown, we will next demonstrate how to calculate indices with the package "vegan." Using this package can be useful when dealing with a larger number of communities.

2.2.1.2 Calculating Alpha Biodiversity Indices with the "Vegan" Packages

First, we install and load the package:

```
#Installing the package "vegan"
install.packages("vegan")
library(vegan)
```

As an example, we will create a matrix with the number of species within four plant communities. Calculation of biodiversity indices will then include creating loops, and some other commands that are useful for more complex data sets.

Here is our matrix:

```
#Number of species - four plant communities
comm.1<-c(25,40,50,15,22,0,0)
comm.2<-c(70,32,58,42,0,2,0)
comm.3<-c(82,50,0,0,24,32,0)
comm.4<-c(0,30,72,75,36,4,47)

#Creating a matrix - four plant communities
bd<-matrix(c(comm.1, comm.2, comm.3, comm.4), nrow=4, byrow = T)
row.names(bd)<-c("comm.1", "comm.2", "comm.3", "comm.4")
colnames(bd)<-paste("Species", LETTERS[1:7])
bd
```

The input data look like this:

	Species A	Species B	Species C	Species D	Species E	Species F	Species G
comm.1	25	40	50	15	22	0	0
comm.2	70	32	58	42	0	2	0
comm.3	82	50	0	0	24	32	0
comm.4	0	30	72	75	36	4	47

Now we can start the calculations. Below, we calculate three biodiversity indices: Richness, Shannon, and Simpson.

2.2.1.2.1   Richness

Richness is an alpha biodiversity index, and it represents the total number of species within an area (Fedor and Zvaríková 2019). In the package "vegan," the index "richness" is not offered, but we can calculate it this way:

```
#Calculating the richness
fun.1<-function(x){sum(x>0)}
richness<-apply(bd, 1, FUN=fun.1)

richness
```

Here are the results:

```
comm.1 comm.2 comm.3 comm.4
     5      5      4      6
```

The results show that we have 5 species in communities: comm.1 and comm.2, 4 species in comm.3, and 6 species in comm.4.

2.2.1.2.2   Shannon Index

To calculate the Shannon index, we use the package "vegan." We create a loop with the command "for:"

```
#Calculating the Shannon index
for (bd.row in 1:4)
{shannon<- matrix(diversity(bd[,], index = "shannon"))}
shannon<-round(shannon,3)
```

```
#Adjusting the names of rows and columns
row.names(shannon)<-row.names(bd)
colnames(shannon)<-"Shannon"
shannon
```

The result is:

```
        Shannon
comm.1   1.522
comm.2   1.386
comm.3   1.278
comm.4   1.601
```

Here, we have calculated the value of the Shannon index for four plant communities using one command. The first plant community is the same as the one from the introductory example, when we calculated this index using a formula, so we can confirm that we have calculated that value correctly.

2.2.1.2.3 Simpson Index

Calculating the value of the Simpson index is similar to the one for calculating the Shannon index.

```
#Calculating the Simpson index
for (bd.row in 1:4)
{simpson<- matrix(diversity(bd[,], index = "simpson"))}
simpson<-round(simpson,3)

#Adjusting the names of rows and columns
row.names(simpson)<-row.names(bd)
colnames(simpson)<-"Simpson"
simpson
```

The result is:

```
        Simpson
comm.1   0.765
comm.2   0.734
comm.3   0.694
comm.4   0.781
```

Now, we can assemble all three indices that we have calculated, and create a data frame:

```
#Putting together all indices
indices<-cbind(shannon, simpson, richness)
indices<-data.frame(indices)
indices
```

And we get:

```
        Shannon Simpson richness
comm.1   1.522   0.765      5
comm.2   1.386   0.734      5
comm.3   1.278   0.694      4
comm.4   1.601   0.781      6
```

We can also plot these results:

```
#Plotting the results
library(ggplot2)
p<-ggplot(data=indices,aes(x=Simpson,y=Shannon,
  label=row.names(indices))) +
  geom_point(aes(color=factor(richness)), size=4) +
  geom_text(hjust=-0.2,vjust=0.1)

#Setting x and y axis range
p<-p+ylim(1.25,1.62)+xlim(0.68, 0.79)

#Changing the name of the legend

p<-p+guides(color=guide_legend(title="Richness"))

#Changing the names of axis
p+xlab("Simpson index")+ylab("Shannon index")
```

The result is presented in Fig. 2.6.

In Fig. 2.6, the value of the Simpson index is placed on the x-axis, the value of the Shannon index is placed on the y-axis, and different colored dots are used to indicate the value of the richness index. The results clearly show that the community labeled comm.4 has the highest values for all three indices, while the community labeled comm.1 has the lowest values for all alpha biodiversity indices we analyzed.

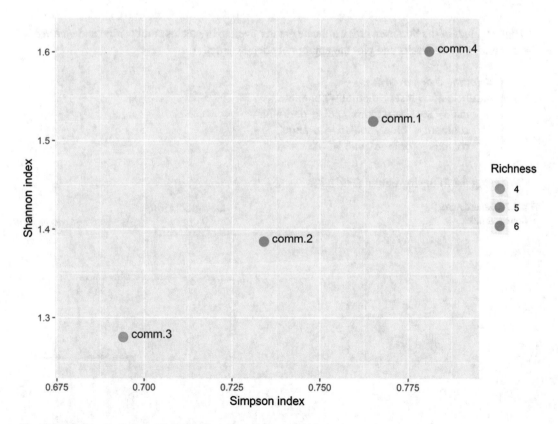

Fig. 2.6 Biodiversity indices (Richness, Shannon and Simpson index)

2.2.1.3 Calculating Beta Biodiversity Indices with the "Vegan" Package

In this section, we calculate two indices of beta diversity with the package "vegan:" the Sorensen index and the Jaccard distance. Both indices are derived numbers (Ma, 2018) measuring similarity (Sorensen index) and dissimilarity (Jaccard distance) between different communities (Begon et al. 1996). More information about these indices can be found with the command:

```
#Help menu for beta indices
?designdist()
```

2.2.1.3.1 Sorensen Index

Following the rule indicated in the help menu, we calculate the Sorensen index:

```
#Calculating Sorensen index
sorensen<-designdist(bd, method="(2*J)/(A+B)", terms=c("binary"))
sorensen<-round(sorensen,3)
sorensen
```

We get the following output:

```
        comm.1 comm.2  comm.3
comm.2  0.800
comm.3  0.667   0.667
comm.4  0.727   0.727   0.600
```

Higher values of the Sorensen index indicate greater overlap in species composition and vice versa. We can cluster these results and plot the appropriate dendrogram.

```
#Plotting a dendrogram
plot(hclust(sorensen, method="complete"),
      main="Sorensen index", col = "darkblue",
      col.main = "blue", col.lab = "orange",
      col.axis = "orange", sub = "")
```

We obtain the following output (Fig. 2.7).

Fig. 2.7 Dendrogram
of Sorensen index

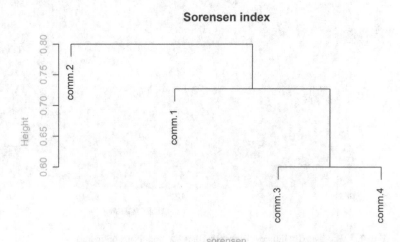

The interpretation of the results is the following: communities labeled as comm.3 and comm.4 differ the most (the value of Sorensen index is 0.6) and the greatest overlap in species composition is between communities comm.1 and comm.2 (the value of Sorensen index is 0.8).

2.2.1.3.2   Jaccard Distance

We now calculate the value of another beta diversity index, the Jaccard distance. The procedure is the following:

```
#Calculating Jaccard distance
jaccard<-vegdist(bd, method="jaccard", binary=T)
jaccard<-round(jaccard,3)
jaccard
```

The results are:

```
        comm.1 comm.2 comm.3
comm.2  0.333
comm.3  0.500    0.500
comm.4  0.429    0.429    0.571
```

These results express differences in species' composition among communities. The interpretation of the results is the following: the value of the Jaccard distance between community 1 and community 2 is 0.333 (which is the lowest value, indicating the greatest overlap in species composition); the Jaccard distance between communities comm.1 and comm.3 is 0.5, and so on. The greatest Jaccard distance is between communities 3 and 4 (the value of 0.571), which indicates the smallest overlap in species' composition occurs between these two communities. These results are similar to those obtained for the Sorensen index.

Our results for the Jaccard distance are presented as the lower triangle of a matrix, but, in the next step, we show how to transform these last results into a complete matrix:

```
#Transforming results into a matrix
jaccard.complete<-stats:::as.matrix.dist(jaccard)
jaccard.complete
```

The result is:

```
        comm.1 comm.2 comm.3 comm.4
comm.1  0.000    0.333    0.500    0.429
comm.2  0.333    0.000    0.500    0.429
comm.3  0.500    0.500    0.000    0.571
comm.4  0.429    0.429    0.571    0.000
```

We added new rows and columns; basically, we created a symmetrical matrix, with the value of zero on the main diagonal. A value of zero means that a community compared to itself has no difference in species' composition.

Now, we can plot the results, but once again we are going to use the function "melt" from the package "reshape":

```
#Melt function
library(reshape)
jaccard.melted<-melt(jaccard.complete)
jaccard.melted
```

Now, we can proceed with plotting the results:

```
#Plotting the results – a base
p<-ggplot(data = jaccard.melted, aes(x=X1, y=X2, fill = value))+
  geom_tile(color = "white")

#Adding gradient scale (white to orange) to the legend
p<-p+scale_fill_gradient2(low = "white", high = "orange",
mid="yellow", midpoint = 0.5, limit = c(0,1),
name="Jaccard distance")

#Adding values of Jaccard distance
p<-p+geom_text(aes(label = value), color = "black", size = 4)

#Removing the labels (x and y axis)
p+xlab("")+ylab("")
```

With the above commands, we have created the output (Fig. 2.8), which makes the results more presentable.

In Fig. 2.8, we have used a gradient scale to display the values of the Jaccard distance, but we can also group the results first, and then plot. If we want to group values of the Jaccard distance, we use the function "cut" to create three groups: the first group will have values ≤ 0.1; the second group will have values >0.1 and $\leq +0.5$, and the third group will have values >0.5. Here is the command:

```
#Cut function
categories.jaccard<-cut(jaccard.melted$value,
breaks=c(-0.01, 0.1, 0.5,1),
labels=c("[0-0.1]", "(0.1-0.5]", "(0.5-1]"))
```

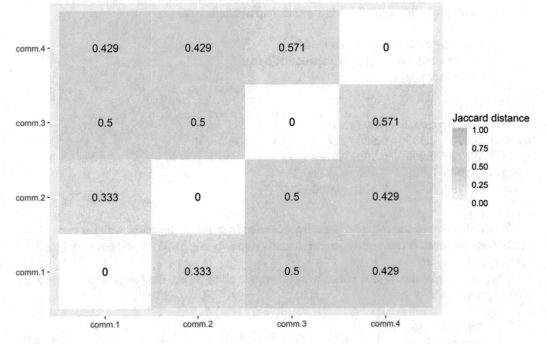

Fig. 2.8 Matrix of Jaccard distance

```
#Adding a new column (categories) to the data frame
jaccard.melted$categories<-categories.jaccard
jaccard.melted
```

Here is the result:

```
      X1       X2   value categories
1  comm.1  comm.1  0.000    [0-0.1]
2  comm.2  comm.1  0.333  (0.1-0.5]
3  comm.3  comm.1  0.500  (0.1-0.5]
4  comm.4  comm.1  0.429  (0.1-0.5]
5  comm.1  comm.2  0.333  (0.1-0.5]
6  comm.2  comm.2  0.000    [0-0.1]
7  comm.3  comm.2  0.500  (0.1-0.5]
8  comm.4  comm.2  0.429  (0.1-0.5]
9  comm.1  comm.3  0.500  (0.1-0.5]
10 comm.2  comm.3  0.500  (0.1-0.5]
11 comm.3  comm.3  0.000    [0-0.1]
12 comm.4  comm.3  0.571    (0.5-1]
13 comm.1  comm.4  0.429  (0.1-0.5]
14 comm.2  comm.4  0.429  (0.1-0.5]
15 comm.3  comm.4  0.571    (0.5-1]
16 comm.4  comm.4  0.000    [0-0.1]
```

Now, we can plot the results:

```
#Plotting the results - a base
p<-ggplot(data = jaccard.melted,
aes(x=X1, y=X2, fill = categories))+geom_tile(color = "white")

#Adding values of Jaccard distance
p<-p+geom_text(aes(label = value),
color = "black", size = 4)

#Changing colors
p<-p+scale_fill_manual(values = c("white", "light green", "darkgreen"))

#Removing the names of x and y axis
p<-p+xlab("")+ylab("")

#Renaming legend
p+labs(fill="Jaccard index")
```

And the output is shown in Fig. 2.9:

2.2.2   Additional Examples

In this section, we show how to perform additional useful operations on data frames. First, we import the excel file containing input data. The name of the file is Trees.xlxs, and it shows the number of species present within a small protected area (a preview of the file is provided in Appendix 2).[2]

---

[2] See Sect. 2.1.1 if you need a reminder on how to import Trees.xlsx in RStudio.

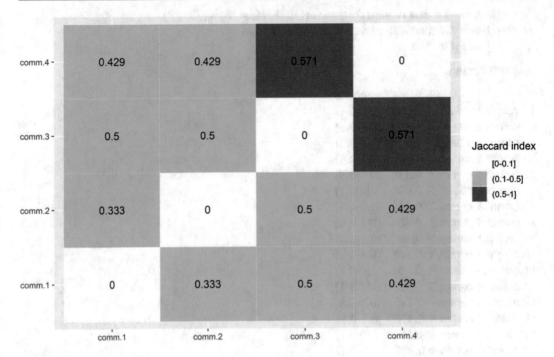

Fig. 2.9 Matrix of Jaccard distance

```
#Importing data file
df.trees<-Trees
df.trees
```

We can check the column names of this data frame:

```
#Checking names of columns
names(df.trees)
```

And we get the output:

```
[1] "Species"    "Height"    "Damages"    "Description"
```

which displays the four column names of the data frame. The column *Species* contains the Latin names of the species; the column *Height* contains their height [ft]; the column *Damages* has two values, true and false, indicating whether the tree has entomological or phytopathological damage (true) or not (false); and the column *Description* has three labels: protected, regular, and invasive (species).

2.2.2.1 Aggregating Values

First, we check the number of trees with and without entomological and phytopathological damage, using the package "dplyr" and the function "count":

```
#Counting the number of damaged trees
library(dplyr)

df.trees%>%count(Damages)
```

And we get:

```
Damages    n
 <chr>   <int>
1 F        40
2 T        16
```

The results show that there are 40 trees without damage, and 16 damaged trees.

A similar procedure can be applied to check how many protected, invasive, and regular species we have:

```
#Counting the number of species by description
df.trees%>%count(Description)
```

And we get:

```
 Description      n
  <chr>        <int>
1 Invasive       24
2 Protected      17
3 Regular        15
```

Now, to combine the two previous counts and make a summary table, we apply the function "xtabs(~)":

```
#Making a summary table (Damages and Description)
table.combined.1<-xtabs(~Damages+Description, data=df.trees)

table.combined.1
```

The result is:

```
         Description
Damages Invasive Protected Regular
      F       14        13      13
      T       10         4       2
```

These last results show how many damaged and undamaged trees there are within each of the three categories (invasive, protected and regular).

In a similar way, we can check the number of damaged and undamaged trees for each species:

```
#Making a summary table (Damages and Species)
table.combined.2<-xtabs(~Species+Damages, data=df.trees)
table.combined.2
```

We can arrange the table, **table.combined.2,** by number of undamaged trees in either ascending or descending order:

```
#Number of undamaged trees – ascending order
table.combined.2[order(table.combined.2[,1]),]

#Number of undamaged trees – descending order
table.combined.2[rev(order(table.combined.2[,1])),]
```

The result is:

Species	Damages F	T
Betula pendula Roth	7	0
Ailanthus altissima (P. Mill.) Swingle	7	1
Platanus x acerifolia (Aiton) Willd.	5	2
Fagus moesiaca (K. Maly) Czecz.	5	2
Robinia pseudoacacia L.	4	2
Populus alba L.	4	0
Quercus robur L.	3	0
Tilia argentea DC.	2	5
Carpinus betulus L.	2	2
Acer negundo L.	1	2

We can turn the values presented in **table.combined.2** into percentages. To do so, we apply "rowSums":

```
#Share of (un)/damaged trees
round(table.combined.2/rowSums(table.combined.2)*100,1)
```

The other option would be to apply:

```
#Share of (un)/damaged trees
fun1<-function(x) {round(prop.table(x)*100,1)}
apply(table.combined.2, 1, FUN=fun1)
```

Here is the result:

Species	Damages F	T
Acer negundo L.	33.3	66.7
Ailanthus altissima (P. Mill.) Swingle	87.5	12.5
Betula pendula Roth	100.0	0.0
Carpinus betulus L.	50.0	50.0
Fagus moesiaca (K. Maly) Czecz.	71.4	28.6
Platanus x acerifolia (Aiton) Willd.	71.4	28.6
Populus alba L.	100.0	0.0
Quercus robur L.	100.0	0.0
Robinia pseudoacacia L.	66.7	33.3
Tilia argentea DC.	28.6	71.4

Now, we can list the name of each species and its description. Because we have duplicate rows (every species has only one label), we can get the result by applying the function "subset" and "!duplicated" as follows:

```
#Removing duplicating rows
desc<-subset(df.trees[-c(2:3)], !duplicated(subset(df.trees,
select = c("Species", "Description"))))
desc
```

In the previous step, we removed columns 2 and 3 ("Height" and "Protected") and here is the output we get:

```
   Species                                Description
   <chr>                                  <chr>
 1 Platanus x acerifolia (Aiton) Willd.   Protected
 2 Ailanthus altissima (P. Mill.) Swingle Invasive
 3 Fagus moesiaca (K. Maly) Czecz.        Protected
 4 Carpinus betulus L.                    Regular
 5 Quercus robur L.                       Protected
 6 Acer negundo L.                        Invasive
 7 Populus alba L.                        Regular
 8 Robinia pseudoacacia L.                Invasive
 9 Tilia argentea DC.                     Invasive
10 Betula pendula Roth                    Regular
```

We will use this data frame at the end of the chapter.

2.2.2.2 Statistical Parameters

Now, we focus on the column "Height." To do some statistical analysis on the values in this column, and calculate the standard deviation, standard error, and confidence interval, the easiest way is to apply the package "Rmisc":

```
#Installing the package "Rmisc"
install.packages("Rmisc")
library(Rmisc)
```

Now, we can apply the function "summarySE" from this package, and we will request statistics for the variable "Height," and for each species separately:

```
#Statistical analysis
stat.df.trees<-summarySE(df.trees,
measurevar = "Height", groupvars = "Species")
stat.df.trees
```

Here is the result:

	Species	N	Height	sd	se	ci
1	Acer negundo L.	3	32.83333	5.879909	3.394767	14.606504
2	Ailanthus altissima (P. Mill.) Swingle	8	50.66250	25.500193	9.015680	21.318694
3	Betula pendula Roth	7	52.25714	4.293739	1.622881	3.971046
4	Carpinus betulus L.	4	39.50000	8.289351	4.144675	13.190207
5	Fagus moesiaca (K. Maly) Czecz.	7	44.25714	9.850018	3.722957	9.109747
6	Platanus x acerifolia (Aiton) Willd.	7	77.11429	10.234977	3.868458	9.465775
7	Populus alba L.	4	81.22500	9.437999	4.719000	15.017963
8	Quercus robur L.	3	74.40000	8.304818	4.794789	20.630311
9	Robinia pseudoacacia L.	6	40.11667	6.915321	2.823168	7.257184
10	Tilia argentea DC.	7	58.07143	4.313434	1.630325	3.989261

We have the results, but the values for *Height*, *sd*, *se*, and *ci* have too many decimal places, so we round them by using the "mutate_at" function from "dplyr" package:

```
#Rounding values - mutate_at funtion
library(dplyr)

stat.df.trees<-stat.df.trees %>%
mutate_at(vars(Height, sd, se, ci), funs(round(., 3)))
stat.df.trees
```

This is the result we get:

Species	N	Height	sd	se	ci
1 Acer negundo L.	3	32.833	5.880	3.395	14.607
2 Ailanthus altissima (P. Mill.) Swingle	8	50.663	25.500	9.016	21.319
3 Betula pendula Roth	7	52.257	4.294	1.623	3.971
4 Carpinus betulus L.	4	39.500	8.289	4.145	13.190
5 Fagus moesiaca (K. Maly) Czecz.	7	44.257	9.850	3.723	9.110
6 Platanus x acerifolia (Aiton) Willd.	7	77.114	10.235	3.868	9.466
7 Populus alba L.	4	81.225	9.438	4.719	15.018
8 Quercus robur L.	3	74.400	8.305	4.795	20.630
9 Robinia pseudoacacia L.	6	40.117	6.915	2.823	7.257
10 Tilia argentea DC.	7	58.071	4.313	1.630	3.989

This result shows, for each species, the number of individuals (N), average height (Height), standard deviation (sd), standard error (se), and width of the 95% confidence interval (ci) about the mean. Now, we can proceed with plotting the results for average height of each species with its standard error bar:

```
#Plotting the results
library(ggplot2)
p<-ggplot(stat.df.trees, aes(x=Species, y=Height))+
    geom_point(aes(col=Species), size=2.8)

#Adding error bar (se)
p<-p+geom_errorbar(aes(ymin=Height-se, ymax=Height+se),
                    width=.2, position=position_dodge(.9))

#Labeling y axis and adjusting text and ticks on x axis
p+ylab("Height [ft]")+
theme(axis.text.x=element_blank(),
        axis.ticks.x=element_blank())
```

The result is presented in Fig. 2.10.

2.2.2.3 Clustering

There are different options for clustering results, and in this section we demonstrate the procedure of clustering height of trees with "k-mean." In the first step, we need to define the number of groups (or classes):

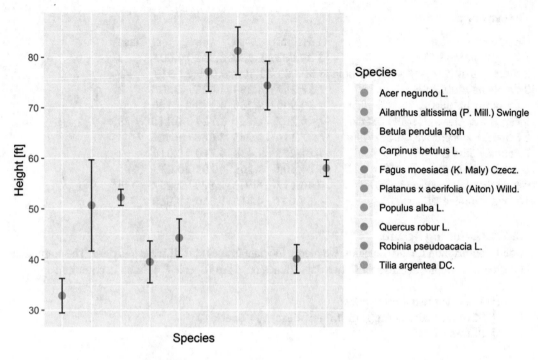

Fig. 2.10   Average height of trees with the standard error bar

```
#Clustering – k means
class<-kmeans(stat.df.trees$Height, 3)
class
```

We get the values, which will be the clustering thresholds:

```
K-means clustering with 3 clusters of sizes 3, 3, 4
Cluster means:
        [,1]
1 77.57976
2 53.66369
3 39.17679
```

Available components:

```
[1] "cluster"  "centers" "totss"  "withinss"
"tot.withinss" "betweenss" "size"
[8] "iter"        "ifault"
```

We use the component "cluster", which shows results as 1, 2, or 3. These values indicate classes of large (1), medium (2), or small (3) trees. We add the "class" column to the data frame **stat.df.trees**:

```
#Adding clustering results to data frame
stat.df.trees$Class<-class$cluster
stat.df.trees
```

The output is:

	Species	N	Height	sd	se	ci	Class
1	Acer negundo L.	3	32.833	5.880	3.395	14.607	3
2	Ailanthus altissima (P. Mill.) Swingle	8	50.663	25.500	9.016	21.319	2
3	Betula pendula Roth	7	52.257	4.294	1.623	3.971	2
4	Carpinus betulus L.	4	39.500	8.289	4.145	13.190	3
5	Fagus moesiaca (K. Maly) Czecz.	7	44.257	9.850	3.723	9.110	3
6	Platanus x acerifolia (Aiton) Willd.	7	77.114	10.235	3.868	9.466	1
7	Populus alba L.	4	81.225	9.438	4.719	15.018	1
8	Quercus robur L.	3	74.400	8.305	4.795	20.630	1
9	Robinia pseudoacacia L.	6	40.117	6.915	2.823	7.257	3
10	Tilia argentea DC.	7	58.071	4.313	1.630	3.989	2

2.2.2.4 Gathering Results

To add a description for each species, we merge the data frames **stat.df.trees** and **desc**. The command in this case is simple, because these two data frames have the "Species" column in common:

```
#Merging two data frames
df.trees.final<-merge(stat.df.trees, desc, by="Species")
df.trees.final
```

We check the average height of trees belonging to each group (protected, invasive, and regular) with the function "aggregate:"

```
#Calculating average height for each group
aggregate(df.trees.final$Height,
by=list(df.trees.final$Description), FUN=mean)
```

The result is:

	Group.1	x
1	Invasive	45.42098
2	Protected	65.25714
3	Regular	57.66071

We see that the largest trees belong to the category *protected*, the shortest to the category *invasive,* and *regular* trees are in between.

We can plot the results and show, for each species, its average height with standard error bar, its description, and class:

```
#Plotting the results
p<-ggplot(df.trees.final, aes(x=Species, y=Height))+
    geom_point(aes(col=Species, shape=factor(Class)), size=2.8)

#Adding error bar (se)
p<-p+geom_errorbar(aes(ymin=Height-se, ymax=Height+se),
                    width=.2, position=position_dodge(.9))

#Adding facet_wrap
p<-p+facet_wrap(~df.trees.final$Description)
```

```
#Labeling y axis and adjusting text and ticks on x axis
p<-p+ylab("Height [ft]")+ theme(axis.text.x=element_blank(),
axis.ticks.x=element_blank())

#Renaming legend
p+labs(shape="Class")
```

The result is presented in Fig. 2.11.

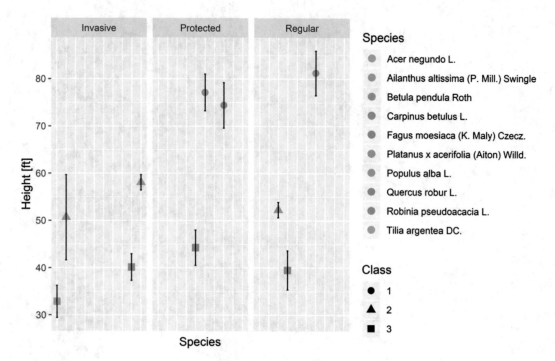

Fig. 2.11 Average height of trees with the standard error bar

Creating Maps

3

In this chapter, we show how to create maps. The chapter is divided in three parts; it deals with mapping on different scales, and explains how to perform the mapping process when input data are: (1) inserted manually or (2) drawn from an R package.

As the first example, we create different maps showing the number of national parks by State in the US. These maps show the number of national parks by using different coloring patterns, but also by treating the number of national parks differently: as an integer, category/group, or as a result of a condition (national park is present—true or not). In addition to creating a map for the entire US, we explain how to create maps when selecting specific States only.

The second example maps the locations of occurrence of two endemic species in Serbia. The example is convenient for showing how to insert the coordinates of locations manually, and then plot them on a map. Some additional features are explained along with the mapping commands, such as clustering the coordinates and creating a geometric figure (ellipse, rectangle) around them.

The third example uses the package "rgbif," which is a comprehensive data base, suited for easy mapping of species occurrence. In this example, instead of inserting coordinates of species occurrence manually, we take them from the package along with other data (country, elevation), and create different types of maps.

3.1   Number of National Parks in the US

In the first example, we create maps presenting the number of national parks in the US by State. We create a data frame with input data and produce different types of maps. Most of the commands we need were already explained in Chap. 1 and in Chap. 2; namely, we use commands such as: "ifelse," "cut," "merge," "subset," etc., but in a slightly different context and these steps are explained in detail.

In this example, we create maps showing the number of national parks by State using continuous and discrete color scale, then the number of national parks is presented as an integer (using the command "as.factor"), then we create a map presenting number of national parks grouped into different categories (using the command "cut"), and finally we create a map showing whether the national parks are present in a State or not (using the command "ifelse"). At the end of the section, we explain how to create maps for certain States, if they are selected by defining a region, or by choosing specific States that we want to present on a map.

Table 3.1 Number of national parks (NP) in the US by state (National Park Service, 2016)

State	Number of NP
Alaska, California	8
Utah	5
Colorado	4
Arizona, Florida, Washington	3
Hawaii, South Dakota, Texas, Wyoming	2
Arkansas, Kentucky, Maine, Michigan, Minnesota, Montana, Nevada, New Mexico, North Carolina, North Dakota, Ohio, Oregon, South Carolina, Tennessee, Virginia	1

Table 3.1 Our input data.

If a State is not present in Table 3.1, it means that it does not have any national parks. For mapping purposes, we use the package "usmap" and "ggplot2." First, we need to install the package "usmap," and load it along with the package "ggplot2:"

```
#Installing the package "usmap"
install.packages("usmap")

#Loading packages
library(usmap)
library(ggplot2)
```

Now, we can get the State names, if we type:

```
#Loading names of states
state.name
state<-state.name
```

We can use State names, and create a column with the number of national parks by using the function "iflese," and the command "or:"

```
#Creating a new column - number of national parks
np.number<-ifelse(state=="Alaska" | state=="California", 8,
ifelse(state=="Utah",5, ifelse(state=="Colorado",4,
ifelse(state=="Arizona"| state=="Florida" | state=="Washington",3,
ifelse(state=="Hawaii" | state=="South Dakota" | state=="Texas" |
state=="Wyoming", 2, ifelse(state=="Arkansas"| state=="Kentucky"|
state=="Maine"| state=="Michigan"| state=="Minnesota"|
state=="Montana"| state=="Nevada"| state=="New Mexico"|
state=="North Carolina"| state=="North Dakota"| state=="Ohio"|
state=="Oregon"| state=="South Carolina"|
state=="Tennessee"| state=="Virginia",1,0))))))

np.number
```

Now, we can assemble a data frame containing our input data:

```
#Creating a data frame - input data
np.df<-data.frame(state, np.number)
np.df
```

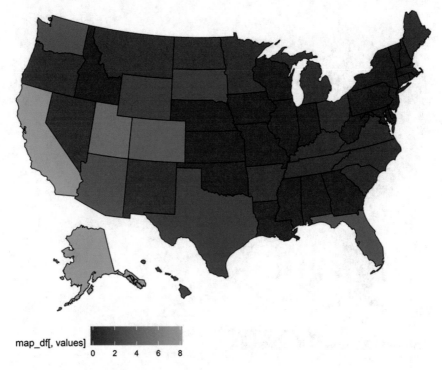

Fig. 3.1 Number of national parks in the US, continuous color scale

and proceed with plotting the map:

```
##Plotting the map – continuous color scale
map.1<-plot_usmap(data=np.df, values="np.number")+
  theme(legend.position="bottom")

map.1
```

Here is the output we get:

Figure 3.1 could be further modified; for example, the color pattern is not very intuitive, because light colors represent higher values and vice versa (this should be the other way around), and the name of the legend also can be changed. Here is how we make the changes:

```
#Adjusting the map
map.1<-map.1+scale_fill_continuous(low="lightgreen",
high="darkgreen")+labs(fill="Number of national parks")
map.1
```

Here is the modified map (Fig. 3.2).

Now, instead of using a continuous color scale, we can use a discrete one. Namely, we can turn the column "np.number," from the data frame **np.df**, into a factor and then plot the results. First, we create a new data frame, a duplicate of the original one, and perform the "as.factor" operation on it:

```
#Creating a duplicate of the data frame
np.df1<-np.df
```

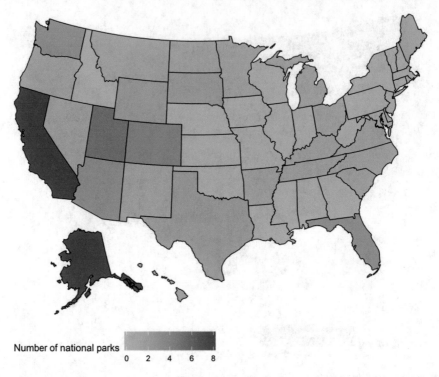

Fig. 3.2 Number of national parks in the US, continuous color scale

```
#Turning the column "np.number" into a factor
np.df1$np.number<-as.factor(np.number)
np.df1
```

Now, we create a map:

```
##Plotting the map – discrete color scale
#Plotting a base map
map.2<-plot_usmap(data=np.df1, values="np.number")

#Adjusting legend name and coloring, removing NA from the legend
map.2<-map.2+scale_fill_brewer(name="Number of national parks",
    palette = 2, na.translate=F)

#Positioning of the legend
map.2<-map.2+theme(legend.position="right")
#Dividing elements of the legend into 2 columns
map.2<-map.2+guides(fill=guide_legend(ncol=2))
map.2
```

Figure 3.3 represents the output.

The number of national parks can be grouped prior to the plotting process. After that, one can create a map with the color scheme matching the rule for grouping the values. First, we group the number of national parks by using the function "cut," and create a new data frame with the results:

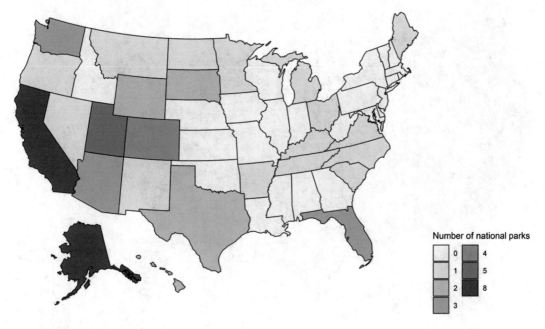

Fig. 3.3 Number of national parks in the US, discrete color scale

```
#Creating groups
groups<-cut(np.df$np.number, breaks=c(-0.1,0,2,4,8),
labels=c("0", "1-2", "3-4", "5-8"))

#Creating a new data frame
np.df2<-cbind(np.df, groups)
np.df2
```

Now, we plot the results:

```
##Plotting the map – grouped results
#Plotting a base map
map.3<-plot_usmap(data=np.df2, values="groups")

#Adjusting legend name and appearance, colors, removing NA
map.3<-map.3+ scale_fill_brewer(name="Number of\nnational parks",
palette = 4, na.translate=F)
#Positioning of the legend
map.3<-map.3+theme(legend.position="right")
map.3
```

Note that here, for dividing the name of the legend into two lines by adding "\n" to the text, we used: "Number of\nnational parks," meaning that "Number of" is the first line and "national parks" is the second line. Figure 3.4 represents the output.

Finally, we can create a map simply showing whether national parks are present or not. Again, we make a duplicate of the original data frame **np.df** and perform the operation we need. In this example, we use the function "ifelse" in the following way: if the number of national parks is greater than 0, we assign "present," and "missing" otherwise.

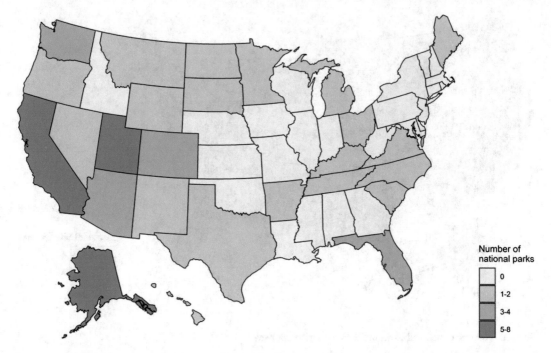

Fig. 3.4 Number of national parks in the US, grouped results

```
#Creating a duplicate of the data frame
np.df3<-np.df

#Testing the presence of national parks
presence<-ifelse(np.df3$np.number>0, "present", "missing")

#Adding the results to the data frame
np.df3$presence<-presence
np.df3
```

We proceed with plotting the results from the data frame **np.df3**.

```
##Plotting the map – based on a condition
#Plotting a base map
map.4<-plot_usmap(data=np.df3, values="presence")

#Adjusting legend name and colors, removing NA from the legend
map.4<-map.4+ scale_fill_manual(values=c("white", "orange"),
name="National parks", na.translate=F)

#Positioning of the legend
map.4<-map.4+theme(legend.position="right")
map.4
```

Figure 3.5 represents the new map, based on a condition (number of national parks greater than 0).

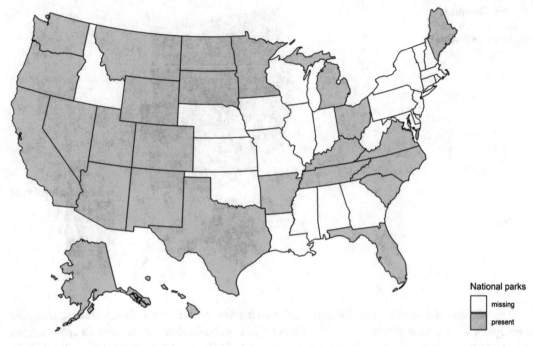

Fig. 3.5 Presence of national parks in the US

3.1.1   Number of National Parks: Selecting States

In this section, we demonstrate the mapping procedure, when the task is to create a map for certain States in the US. We use data frames that we have already created and make appropriate adjustments. We show two cases: the first one is when the program extracts the States based on a region we define; and the second one is when we select (and type) the State names.

First, we create a map for the western region by using a discrete color scheme. We need the data frame named **np.df1**, and the commands look like this:

```
##Plotting a base map –states west region
map.5<-plot_usmap(data=np.df1, values="np.number",
include= .west_region, labels=T)

#Adjusting legend name and coloring, removing NA from the legend
map.5<-map.5+scale_fill_brewer(name="Number of national parks",
palette = 4, na.translate=F)

#Adjusting the legend
map.5<-map.5+theme(legend.position="right")+
guides(fill=guide_legend(ncol=2))
map.5
```

The commands are similar to those for creating the **map.2**. The main difference is in the first command that creates a base map. Here, we selected the western region by applying the command "include= .west_region". In addition to that, we added labels for the States by applying "labels=T". Figure 3.6 is the output.

Fig. 3.6 Number of national parks in the western region of the US

The next task is to create a map for States that we choose ourselves. Therefore, instead of choosing the region, we will type the names (labels) for each State to be included in the map. In this example, we create a map presenting the following states: Oregon, Washington, California, Idaho and Nevada. The number of national parks is defined in the data frame named **np.df2**.

Here are the commands:

```
##Plotting a base map - selected states
map.6<-plot_usmap(data=np.df2, values="groups",
include = c("OR", "WA", "CA", "ID", "NV"), labels = T)

#Final adjustments
map.6<-map.6+scale_fill_brewer(name="Number of national parks",
    palette = 1, na.translate=F)+
theme(legend.position="right")+
guides(fill=guide_legend(ncol=2))
map.6
```

To select the States, we use the command "include=c("OR", "WA", "CA", "ID", "NV")" and add the labels. The result is presented in Fig. 3.7.

In subsequent sections, we mainly deal with mapping assignments on a smaller scale. We show different examples of mapping species occurrence with and without using some of the R packages that support identifying locations of species occurrence.

3.2   Locations of Endemic Species in Serbia

In this section, we demonstrate the procedure of mapping the locations of occurrence of two endemic species in Serbia (*Ramonda serbica* Panč. and *Ramonda nathaliae* Panč. et Petrov.). The input data that we need are the locations' coordinates (longitude and latitude), and we will insert them manually, i.e., without using any R packages containing data regarding species occurrence.

Fig. 3.7 Number of national parks in selected States of the US

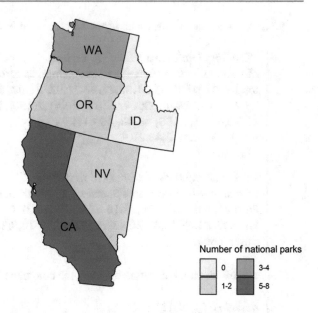

Table 3.2 Locations of *Ramonda serbica* and *Ramonda nathaliae* in Serbia (Rakić et al. 2015)

No.	Species	Long	Lat
1	*Ramonda serbica* Panč.	21.80	44.02
2	*Ramonda serbica* Panč.	21.95	44.02
3	*Ramonda serbica* Panč.	21.80	43.75
4	*Ramonda serbica* Panč.	21.92	43.75
5	*Ramonda serbica* Panč.	21.92	43.65
6	*Ramonda serbica* Panč.	21.92	43.48
7	*Ramonda serbica* Panč.	21.80	43.41
8	*Ramonda serbica* Panč.	22.42	43.20
9	*Ramonda serbica* Panč.	22.16	43.41
10	*Ramonda serbica* Panč.	22.28	43.30
11	*Ramonda serbica* Panč.	20.35	43.12
12	*Ramonda serbica* Panč.	20.45	43.12
13	*Ramonda serbica* Panč.	22.04	43.30
14	*Ramonda serbica* Panč.	22.16	43.30
15	*Ramonda serbica* Panč.	20.41	43.20
16	*Ramonda serbica* Panč.	20.50	43.20
1	*Ramonda nathaliae* Panč. et. Petrov.	22.04	43.20
2	*Ramonda nathaliae* Panč. et. Petrov.	22.16	43.20
3	*Ramonda nathaliae* Panč. et. Petrov.	22.28	43.20
4	*Ramonda nathaliae* Panč. et. Petrov.	22.16	43.13
5	*Ramonda nathaliae* Panč. et. Petrov.	22.04	43.30
6	*Ramonda nathaliae* Panč. et. Petrov.	22.16	43.30
7	*Ramonda nathaliae* Panč. et. Petrov.	22.28	43.13

This procedure of manual input of coordinates is convenient for smaller data sets, and when performing field work. The procedure of creating the map will include creating a data base with input data (coordinates), creating (loading) the map of Serbia, and placing coordinates on the map of Serbia.

Table 3.2 contains the data we need for mapping of locations, and there are 23 locations to be placed on the map (16 for *Ramonda serbica* and 7 for *Ramonda nathaliae*).

First, we create data frames with input data from Table 3.2.

```
#Creating input data frame - R. serbica
df.ramonda<-data.frame(Species="Ramonda serbica Panč.",
long=c(21.80,21.95,21.80,21.92,21.92, 21.92, 21.80,22.42,22.16,
22.28,20.35,20.45,22.04, 22.16, 20.41,20.50), lat=c(44.02,44.02,
43.75,43.75,43.65, 43.48, 43.41,43.20,43.41,43.30,43.12,43.12,
43.30,43.30,43.2,43.2))
df.ramonda

#Creating input data frame - R. nathaliae
df.ramonda1<-data.frame(Species="Ramonda nathaliae Panč. et
Petrov.", long=c(22.04,22.16,22.28,22.16,22.04, 22.16, 22.28),
lat=c(43.20,43.20,43.20,43.13, 43.30,43.30, 43.13))
df.ramonda1
```

Now, we load the map of Serbia by using the command "map_data" and "ggplot2" package:

```
#Loading "ggplot2"
library(ggplot2)

#Loading the world map
world_map <- map_data("world")
ggplot(world_map)
p <- ggplot() + coord_fixed() + xlab("") + ylab("")

#Loading the map of Serbia
serbia_map<-subset(world_map, world_map$region=="Serbia")
serbiaDF<-serbia_map

#Plotting the map of Serbia
serbia.map<-p + geom_polygon(data=serbiaDF, aes(x=long, y=lat,
group=group), fill="lightblue")+xlab("Longitude") +
ylab("Latitude")
serbia.map
```

Now, we place coordinates of *Ramonda serbica* on the map of Serbia using the command "geom_point:"

```
##Plotting the map of R. serbica
ramonda.map1<-serbia.map+geom_point(data=df.ramonda,
aes(x=long, y=lat, color=Species))

#Adjusting colors
ramonda.map1<-ramonda.map1+scale_color_manual(values="red")
ramonda.map1
```

Here, we include the name of species in "aes" by defining "color=Species". In that way, we have specified that species' names should be shown in the legend. Namely, all elements that are placed inside the "aes()" will be placed on the legend as well. That stands for "color", "shape", etc. We use this feature a lot for creating maps in this section.

The map we created in the previous step is presented in Fig. 3.8.

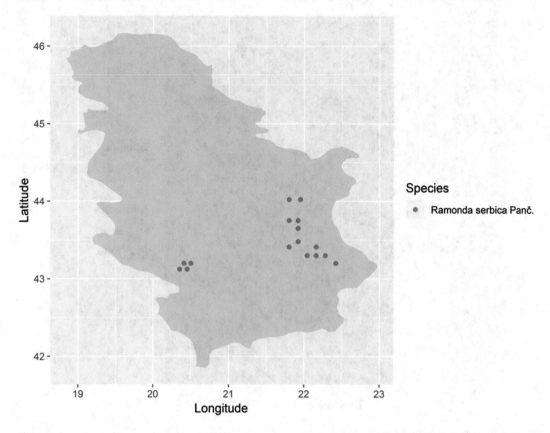

Fig. 3.8 Locations of *Ramonda serbica* in Serbia

Figure 3.8 shows that the locations of *Ramonda serbica* in Serbia are dispersed among two regions, one on the southwest and other the east side of the country. We can cluster locations with the function "kmean" and defining 2 as the number of clustering groups.

```
##Clustering locations - R. serbica
cluster.locs <- kmeans(df.ramonda[2:3], 2)
locations<-cluster.locs$cluster
locations
```

Now, we can create a similar map showing locations as groups on the west and east side, and adding ellipses surrounding locations on each side. For creating an ellipse, we will install and load the package "devtools," and then apply the command "stat_ellipse:"

```
#Installing and loading the package "devtools"
install.packages("devtools")
library(devtools)

##Plotting the map of R. serbica –locations clustered
ramonda.map2<-serbia.map+geom_point(data=df.ramonda, aes(x=long,
y=lat, shape=Species))+ labs(shape="Species")
```

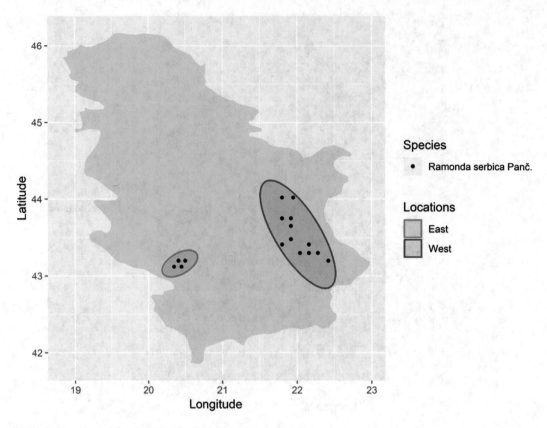

Fig. 3.9 Locations of *Ramonda serbica* in Serbia (clustered)

```
#Adding an ellipse
ramonda.map2<-ramonda.map2+stat_ellipse(data=df.ramonda,
aes(x=long, y=lat, color=factor(locations)),
geom = "polygon", alpha=0.2, size=0.7)
#Additional adjusments
ramonda.map2<-ramonda.map2+
labs(color="Locations")+scale_color_manual(values=c("red", "blue"),
labels=c("East", "West"))

ramonda.map2
```

Figure 3.9 is the output.

Now, we plot the occurrence location for the species *Ramonda nathaliae*. The procedure is similar to the one we used for creating the map of *Ramonda serbica*.

```
##Plotting the map of R. nathaliae
ramonda.map3<-serbia.map+geom_point(data=df.ramonda1,
aes(x=long, y=lat, color=Species))

#Defining colors
ramonda.map3<-ramonda.map3+scale_color_manual(values="dark blue")

ramonda.map3
```

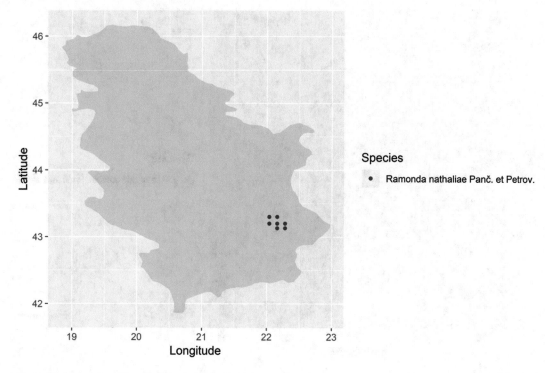

Fig. 3.10 Locations of *Ramonda nathaliae* in Serbia

Figure 3.10 is the output, and a base for creating additional maps.

Now, we can create a geometric figure surrounding the points, i.e., locations presented in Fig. 3.10. If we want to create an ellipse, the procedure is the same as the one we used in the previous example. Here, we decide to create a rectangle, and for that purpose we use the function "geom_rect" from the package "ggplot2". First, we create a rectangle with appropriate values to be placed on the *x* and *y* axis:

```
#Creating a rectangle
rectangle<-data.frame(xmin=min(df.ramonda1$long)-0.1,
xmax=max(df.ramonda1$long)+0.1, ymin=min(df.ramonda1$lat)-0.1,
ymax=max(df.ramonda1$lat)+0.1)

rectangle
```

Here, we have defined that we want to create a rectangle that captures minimal and maximal longitude, and minimal and maximal latitude of locations for *Ramonda nathaliae*, specified in the data frame **df.ramonda1**, and defined 0.1 as distance from each of the points. Now, we place the rectangle on the map named **ramonda.map3**.

```
#Plotting the map of R. nathaliae with "geom_rect"
ramonda.map4<-ramonda.map3+geom_rect(data=rectangle,
aes(xmin=xmin, xmax=xmax, ymin=ymin, ymax=ymax),
color="blue", alpha=0.2)
ramonda.map4
```

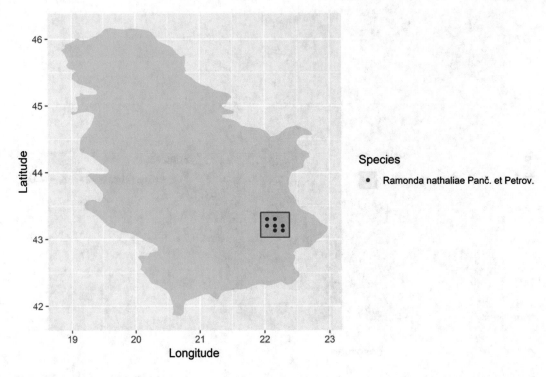

Fig. 3.11 Locations of *Ramonda nathaliae* in Serbia

We have defined that a rectangle should be nearly, to completely, transparent, by defining "alpha=0.2". In that way, the dots (locations) beneath the rectangle are still visible. Figure 3.11 represents the output.

Now, we can inspect whether the locations for these two species overlap. To do so, we apply the command "intersect" from the package "dplyr:"

```
#Loading the package "dplyr"
library(dplyr)

#Checking the overlap of locations
both.ramondas<-intersect(df.ramonda[,2:3], df.ramonda1[,2:3])

#Creating a new data frame
both.ramondas<-data.frame(Species="R. serbica + R. nathaliae",
both.ramondas)
both.ramondas
```

The output we get is:

```
                  Species    long   lat
1 R. serbica + R. nathaliae   22.04  43.3
2 R. serbica + R. nathaliae   22.16  43.3
```

This means that we have two locations where both species occur, and we can check the locations which are unique for *Ramonda serbica*, i.e., where *Ramonda serbica* does and *Ramonda nathaliae* does not exist. The command we use is "setdiff:"

```
#Checking unique locations for R.serbica
r.serbica<-setdiff(df.ramonda[,2:3],df.ramonda1[,2:3])
r.serbica<-data.frame(Species="Ramonda serbica Panč.", r.serbica)
r.serbica
```

Similarly, we can check unique locations for *Ramonda nathaliae,* and then create a data frame with all values: unique locations for each species and mutual ones.

```
#Checking unique locations for R. nathaliae
r.nathaliae<-setdiff(df.ramonda1[,2:3],df.ramonda[,2:3])
r.nathaliae<-data.frame(Species="Ramonda nathaliae Panč. et Petrov.", r.nathaliae)
r.nathaliae

#Creating a new data frame – unique and overlapping locations
ramonda.final<-rbind(r.serbica, r.nathaliae, both.ramondas)
ramonda.final
```

Now, we proceed with plotting the values for the new data frame **ramonda.final**:

```
##Plotting the final map
ramonda.map5<-serbia.map+geom_point(data=ramonda.final,
aes(x=long, y=lat, color=Species),
size=1.5, alpha=0.7)

#Adjusting colors
ramonda.map5<-ramonda.map5+ scale_color_manual(values=c("red", "blue", "purple"))

ramonda.map5
```

Figure 3.12 presents the result, and the locations where the occurrence overlaps are specified with one additional color (purple).

Now, we can present three maps, one next to the other, presenting locations for *Ramonda serbica, Ramonda nathaliae,* and locations where both species are present. We can no longer use the data frame **ramonda.final,** because it would not include the mutual locations when presenting locations of single species. Therefore, we need to create a new data frame containing locations for each of species (including mutual ones), and locations where there is overlap:

```
## Creating a new data frame
ramonda.final1<-rbind(df.ramonda, df.ramonda1, both.ramondas)
```

Note that this data frame **ramonda.final1** has 25 rows, 23 from the original input file plus 2 rows for locations that overlap.

Now, we proceed with creating a map. We use the function "facet_wrap;" in order to create three maps at once, one next to the other; presenting the distribution of the species *R. serbica* and *R. nathaliae,* separately, and then the locations where their distributions overlap.

```
## Plotting the final map - "facet_wrap"
ramonda.map6<-serbia.map+geom_point(data=ramonda.final1,
aes(x=long, y=lat, color=Species),size=1, alpha=0.7)+
  scale_color_manual(values=c("red", "blue", "purple"))
```

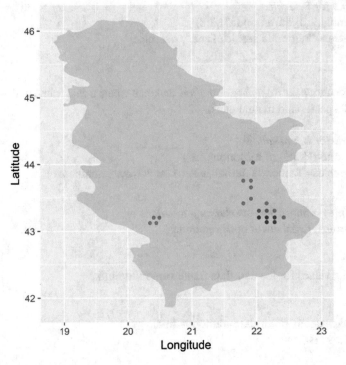

Fig. 3.12 Locations of *Ramonda serbica* and *Ramonda nathaliae* in Serbia

```
#Applying the "facet_wrap" function
ramonda.map6<-ramonda.map6+
facet_wrap(~Species)+theme(legend.position = "none")

ramonda.map6
```

Figure 3.13 is the output.

In this example, we created a data frame with input data, but R offers specialized packages with locations of species. In the next section, we show one of them, called "rgbif." Although using the specialized packages shortens the process of creating maps, sometimes inserting coordinates manually is inevitable, so one should be capable of doing both.

3.3   Package "rgbif"

The R package "rgbif" is a comprehensive data base containing data regarding species occurrence around the world. The acronym "rgbif" stands for R Global Biodiversity Information Facility, and it contains data related to the taxonomy of species, as well as spatial data, i.e., locations of species' occurrence.

In this section, we show how to use this package for mapping purposes. There are some specific commands we need to use, but we also rely on the basic ones, such as "subset," "count," "group_by'" etc.

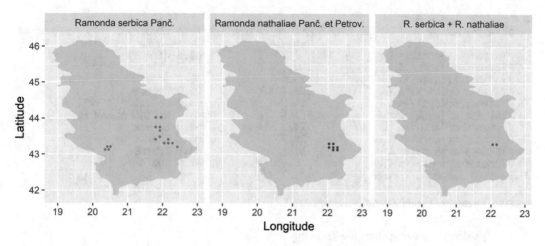

Fig. 3.13 Locations of *Ramonda serbica* and *Ramonda nathaliae* in Serbia

We start by installing the package "rgbif"

```
#Installing the package "rgbif"
install.packages("rgbif")
library(rgbif)
```

First, we demonstrate mapping for one species, and then we include an additional one.

3.3.1   Occurrence of Single Species

With the package installed, we can search for the occurrence of specific species, genera, families, or higher taxonomic classes. For demonstration purposes, we choose to search the occurrence locations for the species *Platanus orientalis* L. If we search for a single species, the structure of the command looks like that presented below, and what needs to be specified is "scientificName:"

```
#Occurrence search for Platanus orientalis
platanus.occ<- occ_search(scientificName = 'Platanus orientalis',
hasCoordinate = TRUE, hasGeospatialIssue = FALSE, limit=10000, start=0)

platanus.occ
```

In the command above, we removed locations from the data base that lack coordinates (hasCoordinate = TRUE), as well as those elements with geospatial issues (hasGeospatialIssue = FALSE). In addition, we have defined that the limit for search is 10,000, in order to obtain a full data base; the default limit is 500, and there usually are more locations where the species has been registered. The data base we obtained contains 145 variables and their names can be inspected if we type:

```
#Inspecting the output data base
platanus.occ$data
```

From the collection of variables, we choose those that are needed for the mapping assignments we have in mind (species' name, longitude, latitude, country, and elevation). Here is the command:

```
# A tibble: 1,599 x 5
   scientificName          decimalLongitude decimalLatitude country          elevation
   <chr>                          <dbl>            <dbl> <chr>                <dbl>
 1 Platanus orientalis L.          7.28            51.8 Germany                 NA
 2 Platanus orientalis L.        109.              34.2 China                   NA
 3 Platanus orientalis L.         25.4             35.2 Greece                  NA
 4 Platanus orientalis L.         25.4             35.2 Greece                  NA
 5 Platanus orientalis L.         24.9             35.0 Greece                  NA
 6 Platanus orientalis L.         21.8             41.7 North Macedonia         NA
 7 Platanus orientalis L.         24.8             40.9 Greece                  NA
 8 Platanus orientalis L.         20.7             38.7 Greece                  NA
 9 Platanus orientalis L.         32.9             34.9 Cyprus                  NA
10 Platanus orientalis L.          2.65            42.2 Spain                   NA
# ... with 1,589 more rows
```

Fig. 3.14 Occurrence locations for *Platanus orientalis* L

```
#Selecting the variables for mapping
platanus<- subset(platanus.occ$data,select=c(scientificName,
decimalLongitude, decimalLatitude, country, elevation))

#Confirming the name of the species
platanus<-subset(platanus,
scientificName=="Platanus orientalis L.")
platanus
```

We get the following output (Fig. 3.14):

Figure 3.14 shows that there are 1599 locations without geospatial issues that are registered within the package "rgbif" as occurrence locations of the species *Platanus orientalis* L. The data bases within the package are constantly being updated, so one can get larger numbers of locations as time goes by.

In the previous step, we obtained coordinates (longitude and latitude) for the occurrence locations, and now we need to place them on the world map. We plot the world map by using the command "map_data" and the package "ggplot2:"

```
#Getting the original data base
world <- map_data("world")

#Plotting the world map
library(ggplot2)
map.world <- ggplot() + coord_fixed() +
xlab("") + ylab("")+ geom_polygon(data=world,
aes(x=long, y=lat, group=group), color="light green",
fill="light green")

#Final adjustments of the world map
map.world<-map.world+theme(axis.ticks=element_blank(),
axis.text.x=element_blank(),
axis.text.y=element_blank())
map.world
```

Now, we place the coordinates from the data frame named **platanus** on the world map. Here are the commands:

```
#Plotting the occurrence locations on the world map
map.platanus.world<- map.world +
geom_point(data=platanus,
```

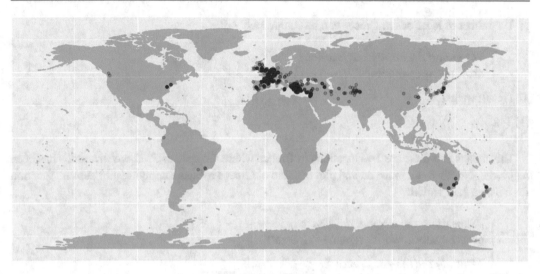

Platanus orientalis L.

Fig. 3.15 Occurrence of *Platanus orientalis*

```
aes(x=decimalLongitude, y=decimalLatitude,
color=scientificName), size=1.2, alpha=0.3)

#Additional fitting
map.platanus.world<-map.platanus.world+
labs(color="")+theme(legend.position="bottom")+
scale_color_manual(values=c("dark blue"))

map.platanus.world
```

Figure 3.15 presents the map we get.

In addition to the map we just created, we can count the number of locations per country where the species has been registered. The command is rather simple; we apply the functions "group_by" and "count" from the package "dplyr:"

```
#Count number of locations per country
library(dplyr)
countries<-platanus%>%
  group_by(country)%>%
  count(country)
countries
```

The result shows that the species is present in 43 counties. Now, we can select one of these countries, and map occurrence locations for that country. Here, we select Greece, because the species is widely spread in Greece. As in earlier examples, the mapping procedure includes three steps: (1) selecting the occurrence locations of *Platanus orientalis* L. in Greece, (2) plotting the map of Greece, and (3) placing the occurrence locations on the map of Greece. Let's start with the first step; it includes subsetting the data frame **platanus** with the condition for name of the country:

```
#Subsetting the original data frame by country (Greece)
platanus.greece<-subset(platanus, country=="Greece")
platanus.greece
```

The nmber of locations in Greece can be counted as:

```
#Counting the number of locations
nrow(platanus.greece)
```

The result we get is:

[1] 142

This means that there are 142 locations in Greece where the species *Platanus orientalis* has been registered. Now, we can move on and plot the map of Greece by again applying the "subset" function for the data base **world**:

```
#Subsetting the original data base
greece<-subset(world, region=="Greece")

#Plotting the map of Greece
map.greece <- ggplot() + coord_fixed() +xlab("Longitude") +
ylab("Latitude")+ geom_polygon(data=greece,aes(x=long, y=lat,
group=group), color=" yellow", fill="yellow")
map.greece
```

Finally, we place locations from the data frame **platanus.greece** on the map of Greece using the function "geom_point:"

```
#Placing locations on the map of Greece
map.platanus.greece<-map.greece+ geom_point(data=platanus.greece,
aes(x=decimalLongitude, y=decimalLatitude, shape=scientificName),
color="dark green",size=1.5, alpha=0.7)

#Renaming the legend
map.platanus.greece<-map.platanus.greece+
labs(shape="Scientific name")
map.platanus.greece
```

Figure 3.16 presents the results.

The data frame **platanus.greece** contains data regarding elevation, and we can use it for producing new maps. Elevation is not known for the all locations, so first we can check the number of missing data:

```
#Counting number of missing elements for elevation
missing.elev<-sum(is.na(platanus.greece$elevation))
missing.elev
```

The result is:

[1] 74

This means that for 142 locations in Greece, there are 74 of them with missing data for elevation. At the moment, elevation data is missing for most species in the package "rgbif," but as we mentioned earlier, the package is constantly being updated.

We plot the map showing the locations where elevation is known and unknown (NA). First, we create a new data frame with a new column containing values: known and NA. We use the package "dplyr" and the commands "group_by" and "ifelse:"

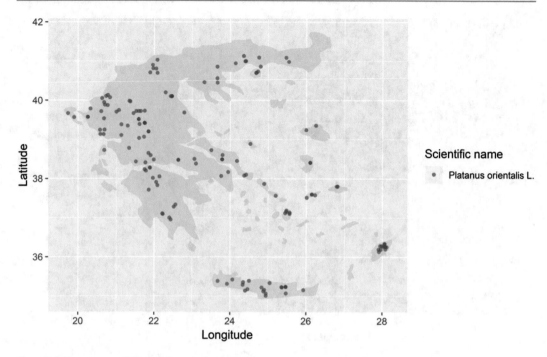

Fig. 3.16   Occurrence of *Platanus orientalis* L. in Greece

```
#Creating a new data frame
library(dplyr)
platanus.greece.1<-platanus.greece %>%
  group_by(Elevation=ifelse(elevation=="NA", NA, "known"))
platanus.greece.1
```

Now we proceed with plotting the map:

```
##Plotting the map – elevation (known, NA)
map.platanus.greece.1<- map.greece +
geom_point(data=platanus.greece.1,
aes(x=decimalLongitude, y=decimalLatitude,
color=Elevation, shape=scientificName), size=1.5, alpha=0.7)

#Additional fitting – Changing colors
map.platanus.greece.1<-map.platanus.greece.1
+scale_color_manual(values="red", na.value="black")

#Additional fitting – Renaming legend (1)
map.platanus.greece.1<-map.platanus.greece1+
  labs(color="Elevation [m]")

#Additional fitting – Renaming legend (2)
map.platanus.greece.1<-map.platanus.greece.1+
labs(shape="Scientific name")

map.platanus.greece.1
```

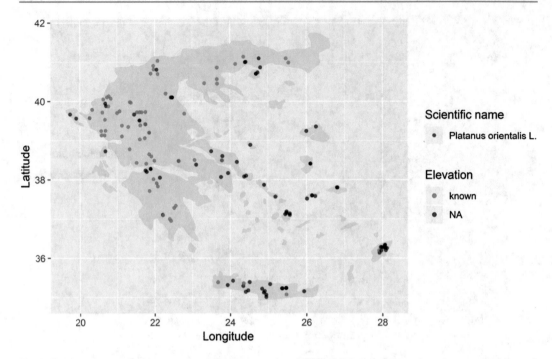

Fig. 3.17 Occurrence of *Platanus orientalis* in Greece and availability of elevation data

Figure 3.17 is the result.

Next, we group elevation in three categories with the function "cut," and proceed with plotting of the results:

```
#Defining elevation groups
Elevation.group<-cut(platanus.greece$elevation,
breaks=c(0,500, 1000, 1500),
labels=c("0-500", "501-1000", "1001-1500"))

#Adding the results to the original data frame
platanus.greece$Elevation.group<-Elevation.group

##Plotting the map – elevation groups
map.platanus.greece.2<- map.greece + geom_point(data=platanus.greece,
aes(x=decimalLongitude, y=decimalLatitude,
color=Elevation.group, shape=scientificName),size=1.5, alpha=0.7)

#Additional fitting – Changing colors
map.platanus.greece.2<-map.platanus.greece.2 +scale_color_manual(values=c("red",
"darkgreen", "blue"), na.value="black")

#Additional fitting – Renaming legends
map.platanus.greece.2<-map.platanus.greece.2+
labs(color="Elevation [m]") + labs(shape="Scientific name")
map.platanus.greece.2
```

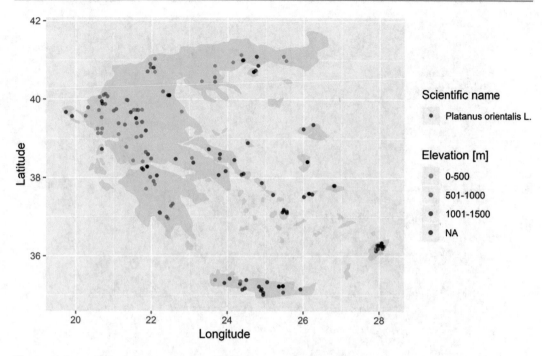

Fig. 3.18 Occurrence of *Platanus orientalis* in Greece with elevation categories

Figure 3.18 is the output.

Along with creating a map containing elevation data, we can inspect elevation, as numeric data, and create a "boxplot" showing the values. First, we need to remove the missing values. Because the missing values only exist in the column "elevation," we can directly apply the function "na.omit" to discard them. After that, we can proceed with plotting the results:

```
#Removing NA values – elevation
platanus.greece.2<-na.omit(platanus.greece)
platanus.greece.2

#Plotting elevation (boxplot)
ggplot(platanus.greece.2, aes(x=scientificName,
y=elevation, fill=country)) +
  geom._boxplot()+ylab("Elevation [m]")+
  xlab("")+labs(fill="Country")
```

Figure 3.19 is the output.

Figure 3.19 can be supported by applying the function "summary," and checking the minimum, maximum, median, mean, first and third quartile for the values of elevation.

```
#Summary function – elevation
summary(platanus.greece$elevation)
```

The result we get is:

Min.	1st Qu.	Median	Mean	3rd Qu.	Max.
5.0	207.2	405.0	451.3	647.5	1350.0

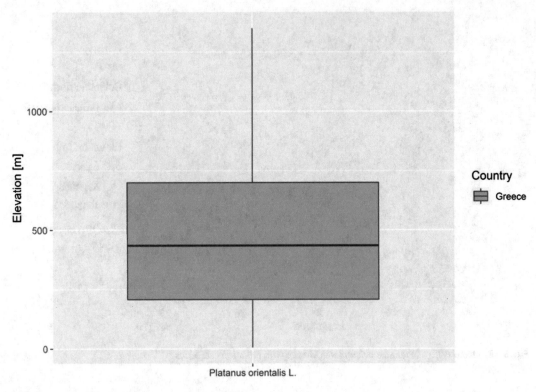

Fig. 3.19 Elevation of *Platanus orientalis* L. in Greece

3.3.2   Occurrence of More than One Species

In this section, we demonstrate how to map the occurrence of different species on the same map. Here, as an additional species, we choose to show the occurrence of *Quercus pubescens* Willd. on the map of Greece. First, we search for the occurrence of this new species to analyze:

```
#Occurrence search for Quercus pubescens – in Greece
quercus.greece.occ<- occ_search(scientificName = 'Quercus
pubescens',hasCoordinate = TRUE,
hasGeospatialIssue = FALSE,
limit=2000, start=0, country="GR")

quercus.greece.occ
```

In the command above, we have defined that we are looking for the locations in Greece specifically, which is why we add "country="GR." Now, we select the variables needed for mapping:

```
#Selecting the variables for mapping
quercus.greece<- subset(quercus.greece.occ$data,
select=c(scientificName, decimalLongitude,
decimalLatitude, elevation))
```

```
#Confirming the name of the species
quercus.greece<-subset(quercus.greece,
scientificName=="Quercus pubescens Willd.")
quercus.greece
```

Next, we can plot the occurrence of *Quercus pubescens* on the map of Greece. The procedure is the same as we used for the previous species (*Platanus orientalis*):

```
#Placing locations on the map of Greece
map.quercus.greece<-map.greece+
geom_point(data=quercus.greece,
aes(x=decimalLongitude, y=decimalLatitude,
shape=scientificName), color="dark blue",
size=1.5, alpha=0.7)

#Renaming the legend
map.quercus.greece<-map.quercus.greece+
labs(shape="Scientific name")
map.quercus.greece
```

Figure 3.20 presents the output.

Next, we create a data frame containing locations of occurrence for both species (*Platanus orientalis* and *Quercus pubescens*). Basically, we need to gather the data frames **platanus.greece** and **quercus.greece**. These two data frames do not have the same structure, because **platanus.greece** has two additional columns named "country" and "Elevation.group," so we need to remove them first, and then proceed with the command "rbind":

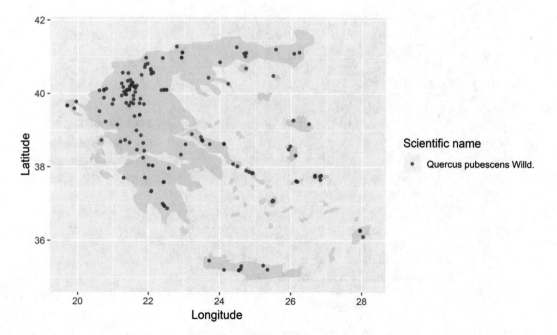

Fig. 3.20 Occurrence of *Quercus pubescens* in Greece

```
#Removing columns "country" and "Elevation.group"
platanus.greece<-platanus.greece[,-c(4,6)]
platanus.greece

#Gathering the data frames
locations<-rbind(platanus.greece, quercus.greece)
locations
```

Now, we have everything set for plotting the first map presenting both species:

```
#Plotting the map combining two species
map.greece.com<-map.greece+ geom_point(data=locations,
aes(x=decimalLongitude, y=decimalLatitude,
color=scientificName),size=1.3, alpha=0.7)

#Additional fitting – Changing colors
map.greece.com<-map.greece.com +
scale_color_manual(values=c("dark green", "dark blue"))

#Renaming the legend
map.greece.com<-map.greece.com+labs(color="Scientific name")

map.greece.com
```

For creating this map, we have defined that species should be presented by different colors, and the rest of the commands are similar to mapping a single species. Figure 3.21 is the output.

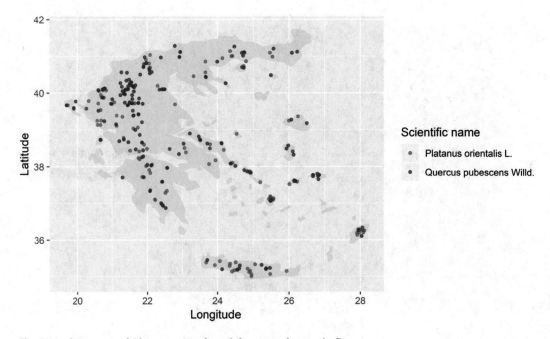

Fig. 3.21 Occurrence of *Platanus orientalis* and *Quercus pubescens* in Greece

Next, we can add a map with elevation categories. First, we modify the data frame **locations** by adding the columns with categories we define by using the function "cut:"

```
#Defining the elevation categories
locations$group<-cut(locations$elevation, breaks=c(0,250, 750, 1500),
labels=c("0-250", "251-750", "751-1500"))

#Plotting the map
map.greece.com.1<-map.greece+ geom_point(data=locations,
aes(x=decimalLongitude, y=decimalLatitude,
shape=scientificName, color=group),
size=1.4, alpha=0.5)

#Additional fitting - Changing colors
map.greece.com.1<-map.greece.com.1
+scale_color_manual(values=c("red", "dark green", "blue"),
na.value="black")

#Renaming the legends
map.greece.com.1<-map.greece.com.1+
labs(color="Elevation [m]")+
labs(shape="Scientific name")

#Defining the order of the legends
map.greece.com.1<-map.greece.com.1+
guides(shape = guide_legend(order=1),
color = guide_legend(order=2))

map.greece.com.1
```

In the commands above, we used shape to indicate the species annotation on the map, and color to distinguish different elevation groups. Also, we previously explained how to define the order of legends to be placed on the map. The output is presented by Fig. 3.22.

Finally, we show the number of occurrence locations for both species on the map of Greece. First, we count the number of locations with the functions from the package "dplyr:"

```
#Counting the number of locations in Greece for both species
library(dplyr)
locations1<-locations%>%
group_by(scientificName)%>%
summarise(count=length(scientificName))
locations1
```

We get the result:

scientificName	count
<chr>	<int>
1 Platanus orientalis L.	142
2 Quercus pubescens Willd.	151

We can place these values on the map of Greece, and in the next step we will specify the exact coordinates where the "count" value is being placed:

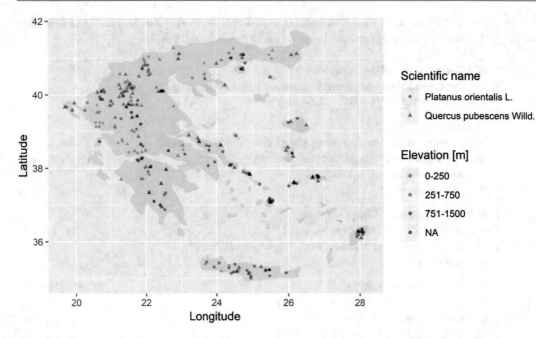

Fig. 3.22 Occurrence of *Platanus orientalis* and *Quercus pubescens* in Greece and elevation groups

```
#Defining place for annotation
locations1$decimalLongitude<-c(21.5,22.1)
locations1$decimalLatitude<-c(39.5,40)
locations1
```

Now, we use the data frame **locations1** to plot the final map:

```
#Plotting the map
map.greece.com.2<-map.greece+geom_point(data=locations1,
aes(x=decimalLongitude, y=decimalLatitude,
color=scientificName), size=10, alpha=0.5)

#Adding text, renaming legend
map.greece.com.2<-map.greece.com.2+ geom_text(data=locations1,
aes(x=decimalLongitude, y= decimalLatitude,
label=count), size=4)+
labs(color="Number of locations")

#Resizing symbols in the legend
map.greece.com.2<-ap.greece.com.2+
scale_color_manual(values=c("dark green", "dark blue"))+
guides(color = guide_legend(override.aes = list(linetype = 0, size=2)))

map.greece.com.2
```

Figure 3.23 is the output, presenting the number of locations for each species.

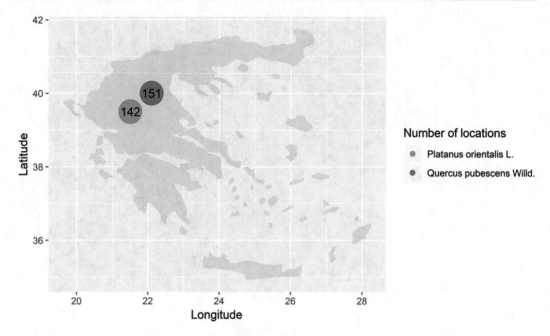

Fig. 3.23 Number of occurrence locations of *Platanus orientalis* and *Quercus pubescens* in Greece

In this chapter, we have demonstrated some of the essential procedures in mapping assignments in ecology. Based on the material we have presented, many similar tasks can be analyzed and resolved.

Basic Statistical Tests

4

Two main groups of statistical tests are parametric and nonparametric tests (Good, 2005). Parametric tests assume that the underlying data are normally distributed and nonparametric tests start off with the assumption that the underlying data do not have a normal distribution (Adler, 2012). Assuming that data are normally distributed is usually correct (Adler, 2012), and therefore parametric tests are more commonly applied than nonparametric ones. Nonparametric tests are also referred to as distribution-free tests, and they can be particularly useful when the data are measured on a categorical scale, i.e., when we deal with an ordered data set (Petrie and Sabin, 2005). It should be noted that nonparametric tests are less powerful in comparison to equivalent parametric tests (Vuković, 1997).

Statistical tests are designed to answer different questions related to the data we analyze (Adler, 2012), and they imply defining a hypothesis. For example, our question might be: Is the population size of a certain species within a particular area equal to 55? In that case, the null hypothesis would be: The population size of the species within a particular area is equal to 55. In contrast to the null hypothesis, there is an alternative hypothesis, and in this case it would be: The population size of the species within a particular area is not equal to 55. From this short example, it can be concluded that the null and alternative hypotheses cover all possible outcomes, meaning that one of them (and only one) must be true.

Given a data set for which the null and alternative hypotheses have been defined, the next steps to test the hypotheses include determining a value of the test statistic and interpreting the p-values and the results (Petrie and Sabin, 2005). The p-value specifies the probability of rejecting the null hypothesis when it is in fact true, and the standard p-value is 0.05 (Vuković, 1997). In subsequent sections, we use the standard value of $p = 0.05$, and explain how to interpret results with respect to the calculated p-value.

In this chapter, we deal with some parametric tests that are most commonly applied for numerical data: independent t-test, dependent t-test, One-Way ANOVA, and Two-Way ANOVA.

4.1 Independent t-test

We apply the independent t-test to check whether there is a statistically significant difference in the mean values of one variable in two different groups (Lalanne and Mesbah, 2016). For example, we apply the t-test if we want to check if there is a statistically significant difference in the sizes of popu-

© Springer Nature Switzerland AG 2020
M. Lakicevic et al., *Introduction to R for Terrestrial Ecology*,
https://doi.org/10.1007/978-3-030-27603-4_4

Table 4.1 Population size at two different altitudes

Location	Population size	
	Altitude 3000 ft	Altitude 5000 ft
1	50	30
2	25	55
3	35	35
4	88	38
5	40	12
6	36	10
7	12	40
8	74	24
9	51	41
10	22	25
11	47	12
12	52	10
13	11	80
14	10	12
15	17	21
16	74	14
17	72	12
18	43	22
19	48	28
20		32

lations of the same species at two different altitudes. Let's say that we have estimated or measured the sizes of populations of the same species (in the same measurement unit) at two different altitudes: 3000 and 5000 ft. (Table 4.1), and that we want to test whether the mean values of the two population sizes differ.

Note that the size of the samples for the two groups is not the same in Table 4.1; for the attitude of 3000 ft, the population size is 19, while and for the altitude of 5000 ft it is 20. Unequal size of samples is not a constraint for applying the t-test.

First, we create the input data in R:

```
#Population 1
population1<-c(50,25,35,88,40,36,12,74,51,22,47,52,
              11,10,17,74,72,43,48)

#Population 2
population2<-c(30,55,35,38,12,10,40,24,41,25,
              12,10,80,12,21,14,12,22,28,32)
```

In the next step, we test whether the variances of **population1** and **population2** differ, and for that purpose we apply the "F test". For the F test, the null hypothesis (H_0) is that the variances are equal, and the alternative hypothesis (H_1) is that they differ. The procedure is the following:

```
#Test to compare two variances
var.test(population1,population2)
```

Here is the output:

F test to compare two variances

data: population1 and population2
F = 1.7301, num df = 18, denom df = 19, p-value = 0.245
alternative hypothesis: true ratio of variances is not equal to 1
95 percent confidence interval:
  0.6796059 4.4574161
sample estimates:
ratio of variances
        1.730078

The results show that the p-value is 0.245, and because $0.245 > 0.05$ it means that, at the significance level of 0.05, we fail to reject the null hypothesis, and conclude that the variances of **population1** and **population2** are equal. Therefore, we can proceed with the t-test:

```
#t-test
t.test(population1, population2, var.equal = T)
```

Note that, if the F test demonstrated that the variances are unequal, we could proceed with the t-test, but in that case the command would look like this: "t.test(population1, population2, var. equal = F)."

For the t-test we perform, the null hypothesis (H_0) is that mean values of **population1** and **population2** are equal, meaning that there is no significant statistical difference between them, and the alternative hypothesis (H_1) is that they differ. The output of the t-test is presented below:

Two Sample t-test

data: population1 and population2
t = 2.2574, df = 37, p-value = 0.02997
alternative hypothesis: true difference in means is not equal to 0
95 percent confidence interval:
  1.518536 28.128832
sample estimates:
mean of x mean of y
 42.47368 27.65000

Based on the results, the p-value of is 0.029, which is smaller than the significance level of 0.05, indicating that the null hypothesis (H_0) should be rejected, and the alternative hypothesis (H_1) is accepted. The alternative hypothesis states that there is a significant difference between mean values of **population1** and **population2**. The results also show the mean values: 42.47 for **population1** and 27.65 for **population2**.

In terms of further interpreting the results, we could say that we expect the size of population of this species to decrease with increase in altitude, if all other factors that might affect the size of population are considered constant.

4.2   Dependent t-test

The dependent t-test is applied when we want to check whether there is a statistically significant difference in mean values of one variable in two related groups (Lalanne and Mesbah, 2016). For example, we can use the dependent t-test if to check whether there is a difference in mean values of a variable in two repeated measurements, or to check the difference in mean values of a variable within the same group in two different periods, for example before and after certain treatment.

In this section, we demonstrate how to apply the dependent t-test on an example related to estimating the erosion rate for the same locations with two different models (Table 4.2). In this example, the main purpose of applying the t-test is checking whether Model A and Model B estimate statistically significant different values of erosion rates.

We start with creating the input data from Table 4.2.

```
#Input data - model A
model.a<-c(0.52, 0.63, 1.27, 1.58, 0.68, 0.38,
0.45, 1.50, 0.54,0.48, 1.21, 1.52, 1.25, 0.53,
1.24, 0.84, 0.37, 1.12)
#Input data - model B
model.b<-c(0.54, 0.64, 1.23, 1.59, 0.7, 0.41,
0.45, 1.57, 0.58, 0.47, 1.24, 1.61, 1.35,
0.56, 1.28, 0.92, 0.39, 1.21)
```

Next, we compare the variances for **model.a** and **model.b**. To compare the variances, we apply the F test as we did in Sect. 4.1, with the null hypothesis (H_0) stating that variances are equal and, the alternative hypothesis (H_1) stating that there is a statistically significant difference between the variances. Here is the command:

Table 4.2 Erosion rate [mm/year]

Location	Erosion rate [mm/year]	
	Model A	Model B
1	0.52	0.54
2	0.63	0.64
3	1.27	1.23
4	1.58	1.59
5	0.68	0.70
6	0.38	0.41
7	0.45	0.45
8	1.50	1.57
9	0.54	0.58
10	0.48	0.47
11	1.21	1.24
12	1.52	1.61
13	1.25	1.35
14	0.53	0.56
15	1.24	1.28
16	0.84	0.92
17	0.37	0.39
18	1.12	1.21

```
#Test to compare two variances
var.test(model.a,model.b)
```

Here is the result of the F test:

F test to compare two variances

data: model.a and model.b
F = 0.93231, num df = 17, denom df = 17, p-value = 0.8868
alternative hypothesis: true ratio of variances is not equal to 1
95 percent confidence interval:
  0.348748 2.492340
sample estimates:
ratio of variances
      0.9323081

The result shows that the p-value is 0.89, therefore the null hypothesis (stating that the variances are equal) cannot be rejected at the significance level of 0.05, so we can proceed with the t-test. The command is similar to the one from Sect. 4.1, but we add "paired = T", because here we are dealing with two related groups:

```
#t-test (paired)
t.test(model.a, model.b, paired = T)
```

Here is the result:

Paired t-test

data: model.a and model.b
t = -3.9155, df = 17, p-value = 0.001114
alternative hypothesis: true difference in means is not equal to 0
95 percent confidence interval:
  -0.0538592 -0.0161408
sample estimates:
mean of the differences
          -0.035

The results show that the difference in mean values of erosion rates, estimated by the models A and B, is −0.035 mm/year. This difference is statistically significant, because the associated p-value is 0.001, which is much smaller than the standard significance level of 0.05, so we reject the null hypothesis in this case. The main conclusion is that Model A and Model B are expected to provide different results for erosion rates. Notice that the 95 percent confidence interval for mean differences does not include 0.

4.3   One-Way ANOVA

ANOVA (Analysis of Variance) is applied when we analyze more than two groups (Petrie and Sabin, 2005). ANOVA compares variability across groups with variability within groups, and if variability among groups exceeds variability within groups, then the conclusion is that at least one group is statistically different from the others (Smalheiser, 2017).

Table 4.3 Seed germination [%] under different light regimes

No.	Germination [%] – Habitat A		
	24 h light	24 h dark	12 h light/12 h dark
1	82	71	81
2	80	73	80
3	82	72	81
4	84	71	78
5	79	72	79
6	81	68	80
7	79	69	80
8	83	72	81
9	82	71	82
10	83	72	79
11	79	72	82
12	81	71	81
13	81	73	80
14	81	68	80
15	80	71	79
16	82	70	81
17	81	72	81
18	83	70	80
19	82	72	78
20	80	72	82

One-Way ANOVA is based on the assumption that the differences between mean values of groups are related to a single factor, and that factor can consist of several levels (Vuković, 1997). Basically, One-Way ANOVA includes one dependent variable and one independent variable—that is, a factor with three or more levels. An example dataset is provided in Table 4.3. For this example, the research question is: Are there statistically significant differences in mean values of seed germination under different light regimes? Here, seed germination is a dependent variable, and light regime is an independent variable with three levels (24 h light, 24 h dark and 12 h light/12 h dark). We will assume that the seed has been collected from a semi-arid type of habitat (this is important for the analysis in a subsequent section).

We create a data frame with the input data:

```
#Creating a data frame - regime=light
light<-data.frame(Light.Regime="Light",
Germination=c(82,80,82,84,
79,81,79,83,82,83,79,81,81,81,80,82,81,83,82,80))

#Creating a data frame - regime=dark
dark<-data.frame(Light.Regime="Dark", Germination=c(71, 73,
72,71, 72,68, 69, 72,71,72,72,71,73,68,71,70,72,70,72,72))

#Creating a data frame - regime=light/dark
light.dark<-data.frame(Light.Regime="Light/Dark",
Germination=c(81,80,81,78,79,80,80,81,82,79,
82,81,80,80,79,81,81,80,78,82))

#Creating a complete data frame
lab.results<-rbind(light, dark, light.dark)
lab.results
```

Before starting with statistical testing, we can get acquainted with the input data, and plot the values of seed germination with respect to different light regimes by using the "geom_boxplot" function:

```
#Plotting the values - boxplot
library(ggplot2)
ggplot(lab.results)+ geom_boxplot(aes(x=Light.Regime,
y=as.numeric(Germination)))+ylab("Germination [%]")+
xlab("Light regime")
```

Figure 4.1 is the output.

Based on Fig. 4.1, seed germination has the lowest values for the light regime "dark." If we examine the values of seed germination for the regimes, "light" and "light/dark," we might be tempted to conclude that the values are higher for the regime "light", but we need a statistical test to check whether these differences are statistically significant.

As mentioned at the beginning of this section, if there are more than two groups to compare, we apply the ANOVA test. Before applying this test, we need to check the homogeneity of variances for all groups. We apply Levene's test for checking the homogeneity of variances within groups. Levene's test is an equivalent for the F test that is used prior to the t-test.

First, we need to install and load the package "car" and then proceed with Levene's test:

```
#Installing the package "car"
install.packages("car")
library(car)

#Checking the homogeneity of variance
leveneTest(Germination~Light.Regime, data=lab.results)
```

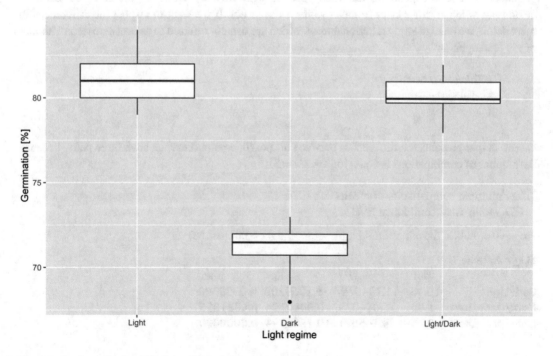

Fig. 4.1 Seed germination [%] under different light regimes

The output we get is presented below:

```
Levene's  Test for Homogeneity of Variance (center = median)
      Df  F value  Pr(>F)
group  2  0.2788  0.7577
      57
```

Similar to before, the *p*-value of 0.757 indicates that we cannot reject the null hypothesis at the significance level of 0.05, so variances are equal across all groups, and therefore we can proceed with the ANOVA test:

```
#ANOVA test
anova<-aov(Germination~Light.Regime, data=lab.results)
summary(anova)
```

Note that the structure of the command is "aov(dependent variable ~ independent variable, data = input data frame)." Here is the output:

```
              Df  Sum Sq  Mean Sq  F value   Pr(>F)
Light.Regime   2  1251.6   625.8    332.4   <2e-16 ***
Residuals     57   107.3     1.9
---
Signif. codes:  0 '***' 0.001 '**' 0.01 '*' 0.05 '.' 0.1 ' ' 1
```

The result shows that the *p*-value is <<0.05, therefore the null hypothesis should be rejected. This means that, according to the ANOVA test, there are statistically significant differences in the germination of seed exposed to the three different light regimes. However, the ANOVA test only determines whether differences in mean values exist or not. If we want to inspect the differences in more detail, we need to apply additional tests, which are usually referred to as post-hoc tests. The first one is "TukeyHSD:"

```
#TukeyHSD test
t<-TukeyHSD(anova)
t
```

The output presents the differences between all possible pairs (here we have three pairs, because the number of combinations is equal to: $3 \times 2/2 = 3$):

```
Tukey multiple comparisons of means
  95% family-wise confidence level

Fit: aov(formula = Germination ~ Light.Regime, data = lab.results)

$Light.Regime
                  diff         lwr          upr       p adj
Dark-Light       -10.15 -11.194079  -9.1059206  0.0000000
Light/Dark-Light  -1.00  -2.044079   0.0440794  0.0631856
Light/Dark-Dark    9.15   8.105921  10.1940794  0.0000000
```

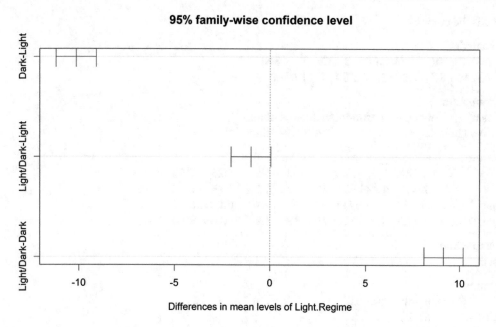

Fig. 4.2 Tukey test—graphical presentation

These values can be presented graphically. The command is simple:

```
#TukeyHSD test - plotting the values
plot(t, col="blue")
```

Figure 4.2 is the output.

Figure 4.2 shows the differences in mean levels for all three pairs of light regimes with a confidence interval of 95% (standard value). Based on the figure, it is easy to conclude that the smallest difference in mean levels is for the pair "light/dark" and "light", and the largest for the pair of regimes "dark" and "light." The results can be inspected by applying another statistical test, the "HSD test" (Honest Significant Difference) from the package "agricolae."

First, we install the package "agricolae," and load the command for the HSD test. The structure for the HSD test is: "HSD.test(anova results, "levels of independent variable", group=T)." The main purpose of the HSD test is to classify the results into appropriate homogenous groups. Homogeneous groups are labeled using letters, and if two (or more) levels of a factor (independent variable) are labeled with the same letter, then there is no statistical difference between these levels.

```
#Installing and loading the package "agricolae"
install.packages("agricolae")
library(agricolae)

#HSD test
posthoc1 <-HSD.test(anova,"Light.Regime", group=T)
posthoc1
```

Here is the output:

```
$`statistics`
   MSerror  Df      Mean        CV       MSD
 1.882456  57  77.53333  1.769595  1.044079

$parameters
    test          name.t ntr StudentizedRange alpha
  Tukey Light.Regime    3           3.403189  0.05

$means
            Germination        std   r  Min  Max    Q25   Q50  Q75
Dark             71.10  1.447321  20   68   73  70.75  71.5   72
Light            81.25  1.446411  20   79   84  80.00  81.0   82
Light/Dark       80.25  1.208522  20   78   82  79.75  80.0   81

$comparison
NULL

$groups
            Germination  groups
Light            81.25       a
Light/Dark       80.25       a
Dark             71.10       b

attr(,"class")
[1] "group"
```

In the previous step, we have determined homogeneous groups. We can add these results to the original data frame, **lab.results,** with the input data. The results are displayed as a new column. Here is the procedure:

```
#Adding the results (groups)
lab.results1<-lab.results
lab.results1$posthoc[lab.results1$Light.Regime=="Light"|
lab.results1$Light.Regime=="Light/Dark"]<-"a"
lab.results1$posthoc[lab.results1$Light.Regime=="Dark"]<-"b"
lab.results1
```

Now, we can plot the results:

```
#Plotting the results
library(ggplot2)
p1<-ggplot(data=lab.results1, aes(x=Light.Regime,
y=as.numeric(Germination)))+
  geom_boxplot(aes(fill=lab.results1$posthoc))

#Additional adjustments
p1<-p1+ylab("Germination [%]")+xlab("Light regime")+labs(fill="Groups")
p1
```

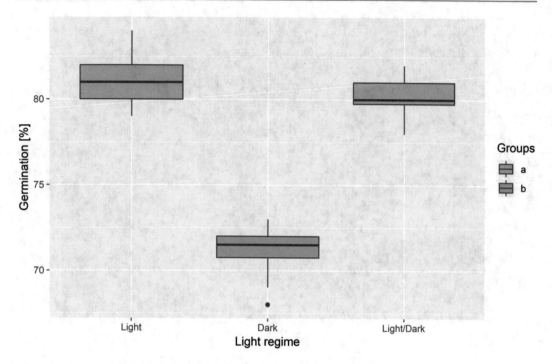

Fig. 4.3 Seed germination [%] under different light regimes—Tukey HSD test

Figure 4.3 presents the output.

Based on the results presented in Fig. 4.3, there is no statistically significant difference in mean values of seed germination under the light regimes labeled as "light" and "light/dark", when taking into account the standard level of significance. Namely, the highest values of germination are recorded for the light regimes "light" and "light/dark," and they are not statistically different, and the lowest values are recorded for the regime labeled as "dark."

Figure 4.3 could be further modified; we can order the values in ascending order by applying the command "reorder:"

```
#Reordering the results
p2<-ggplot(data=lab.results1,
aes(x = reorder(Light.Regime, as.numeric(Germination),
  FUN = median), y=as.numeric(Germination)))+
  geom_boxplot(aes(fill=lab.results1$posthoc))+
  ylab("Germination [%]")+xlab("Light regime")+
  labs(fill="Groups")

p2
```

Figure 4.4 is the final output in this section.

It should be noted that the arcsine-square root transformation as a way to deal with percentage (or proportion) data in ANOVA and other parametric statistics, but, in this particular case, it did not help conform the example data to normality.

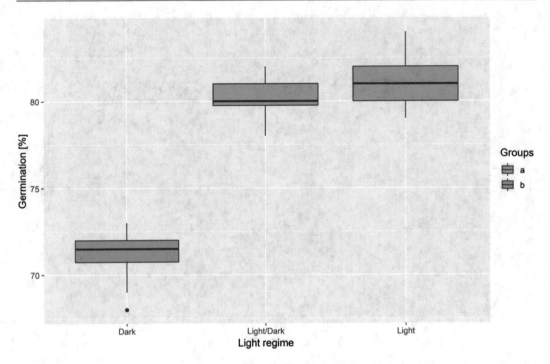

Fig. 4.4   Seed germination [%] under different light regimes—Tukey HSD test

4.4   Two-Way ANOVA

If we extend the One-Way ANOVA with one additional factor, we get a Two-Way ANOVA test (Kéry, 2010). The Two-Way ANOVA is based on the assumption that the differences in the mean values in groups are the consequence of two factors, and again these factors can consist of several levels (Vuković, 1997). In the case of the Two-Way ANOVA, we analyze one dependent variable and two independent variables—factors with at least two levels.

Let's expand the example from the previous section, and suppose that we have collected the seed of the same species from another type of habitat—a humid one, and that we want to introduce newly collected material in the analysis. In this case, we can apply the Two-Way ANOVA test. The input data we need is germination of the seed, collected from the new type of habitat under the same light regimes. Here, germination of the seed is the dependent variable and the two factors are: (1) Habitat, with two levels, *Habitat A,* representing a semi-arid habitat type and *Habitat B,* representing a humid habitat type; and (2) Light regime, with three levels, 24 h light, 24 h dark and 12 h light/12 h dark.

In this example, Two-Way ANOVA answers the following questions:

1. Is there a statistically significant difference in the mean values of seed germination depending on the type of habitat?
2. Is there a statistically significant difference in the mean values of seed germination depending on the light regime?
3. Is there a statistically significant difference in the mean values of seed germination depending on the interaction of the type of habitat and the light regime?

The Two-Way ANOVA provides answers in the "yes/no" manner discussed for the One-Way ANOVA, and we apply post-hoc tests for more detailed analysis as before. Table 4.4 presents our input data.

Table 4.4 Seed germination [%] under different light regimes

No.	Germination [%] – Habitat B		
	24 h light	24 h dark	12 h light/12 h dark
1	72	58	68
2	71	59	70
3	70	62	72
4	69	58	71
5	72	61	72
6	68	58	70
7	70	59	68
8	68	61	71
9	70	60	72
10	73	62	73
11	70	61	70
12	72	59	70
13	71	61	71
14	72	58	72
15	72	61	69
16	73	59	68
17	72	60	70
18	72	61	70
19	70	62	68
20	72	60	69

We start off by creating the input data frame:

```
##Habitat B
#Creating a data frame - regime=light
light1<-data.frame(Light.Regime="Light",
Germination=c(72,71,70,69,72,68,70,
68,70,73,70,72,71,72,72,73,72,70,72,70))
#Creating a data frame - regime=dark
dark1<-data.frame(Light.Regime="Dark",
 Germination=c(58,59,62,58,61,58,59,61,60,62,
61,59,61,58,61,59,60,61,62,60))

#Creating a data frame - regime=light/dark
light.dark1<-data.frame(Light.Regime="Light/Dark",
Germination=c(68,70,72,71,72,70,68,71,72,73,
70,70,71,72,69,68,70,70,68,69))

#Creating a complete data frame
lab.results1<-rbind(light1, dark1, light.dark1)
lab.results1
```

Next, we merge that data for both habitats (Habitat A and Habitat B):

```
#Merging the results - Habitat A and B
lab.results<-data.frame(Habitat="Habitat A", lab.results)
lab.results1<-data.frame(Habitat="Habitat B", lab.results1)
lab.results.final<-rbind(lab.results, lab.results1)
lab.results.final
```

Now, we have everything set for performing the analysis. Before starting the Two-Way ANOVA, we will inspect the homogeneity of variances in all groups. Once again, as in case of the One-Way ANOVA, we apply Levene's test:

```
#Checking homogeneity of the variance
leveneTest(Germination~as.factor(Habitat)*as.factor(Light.Regime),
        data=lab.results.final)
```

Here is the output.

```
Levene's Test for Homogeneity of Variance (center = median)
      Df  F value  Pr(>F)
group  5  0.3157  0.9027
     114
```

Levene's test indicates that the variances are equal across groups (p-value is $0.9 > 0.05$), therefore we can proceed with the Two-Way ANOVA. Here are the commands:

```
#Two-Way ANOVA
anova2<-aov(Germination~as.factor(Habitat)*as.factor(Light.Regime),
data=lab.results.final)
```

At this point, we need to check one more condition related to normality of the distribution of residuals. First, we create a histogram of the residuals, and then overlay it with a curve of the normal distribution:

```
#Histogram of residuals
hist(anova2$residuals, freq=F, breaks=8, ylim=range(0,0.35))
#Normal curve
lines(seq(-4, 3, by=0.25), dnorm(seq(-4, 3, by=0.25),
mean(anova2$residuals), sd(anova2$residuals)), col="red")
```

The output is presented in Fig. 4.5. The normality of the distribution can be checked in other ways, for example, by applying statistical tests such as the Shapiro-Wilk or Anderson-Darling normality tests.

Based on Fig. 4.5, however, the residuals have an approximately normal distribution, so the requirements for proceeding with the Two-Way ANOVA test are fulfilled, and we can calculate the results:

```
#Results – Two-Way ANOVA
summary(anova2)
```

Here are the results:

	Df	Sum Sq	Mean Sq	F value	Pr(>F)	
as.factor(Habitat)	1	3318	3318	1623.055	<2e-16	***
as.factor(Light.Regime)	2	2727	1364	667.014	<2e-16	***
as.factor(Habitat):as.factor(Light.Regime)	2	6	3	1.398	0.251	
Residuals	114	233	2			

Signif. codes: 0 '***' 0.001 '**' 0.01 '*' 0.05 '.' 0.1 ' ' 1

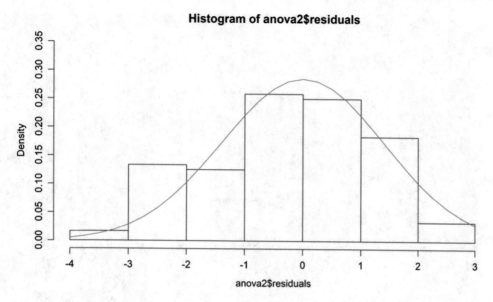

Fig. 4.5 Histogram of residuals

The results show that there are statistically significant differences in mean values of seed germination for both factors separately (Habitat and Light Regime), and that there is no statistically significant difference for their interaction (*p*-value is equal to 0.251). Referring back to the beginning of this section, the answers to the questions are: (1) Yes, (2) Yes and (3) No. This basically means that the impact of light regime can be generalized for both types of habitats.

In order to provide a more detailed analysis, we again apply post-hoc tests, and we select the TukeyHSD test:

```
#Tukey test
t2<-TukeyHSD(anova2)
```

Now, we can inspect the results of the TukeyHSD test for each factor separately, and then for their interaction. We start with the factor "Habitat" and then for the factor "Light regime:"

```
#Tukey test – factor "Habitat"
t2$`as.factor(Habitat)`

#Tukey test – factor "Regime"
t2$`as.factor(Light.Regime)
```

Here are the results.

```
#Tukey test – factor "Habitat"
> t2$`as.factor(Habitat)`
                        diff       lwr        upr       p adj
Habitat B-Habitat A  -10.51667  -11.03379  -9.999543  2.431388e-14

> #Tukey test – factor "Regime"
> t2$`as.factor(Light.Regime)`
```

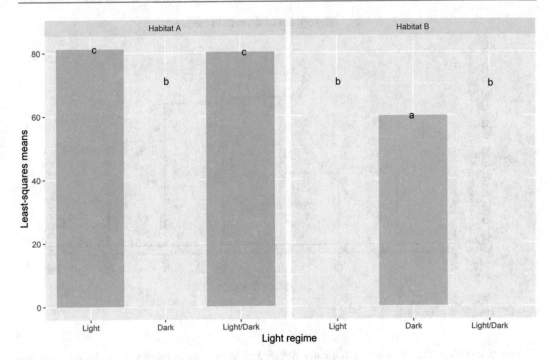

Fig. 4.6 Seed germination [%], different light regimes and types of habitat

	diff	lwr	upr	p adj
Dark-Light	-10.500	-11.259222	-9.74077823	2.431388e-14
Light/Dark-Light	-0.825	-1.584222	-0.06577823	2.974202e-02
Light/Dark-Dark	9.675	8.915778	10.43422177	2.431388e-14

Based on the results, the mean values of germination depend on the type of habitat, and seed germination is statistically higher for the Habitat A (the difference is equal to 10.52%). Similarly, the mean values of germination depend on the light regime and Fig. 4.6 presents differences in mean values for the three light regime pairs.

Now, we can inspect whether the mean values of germination depend on the interaction between two factors, Habitat and Light Regime. Here is the command:

```
#Tukey test – interaction of factors
d<-t2$`as.factor(Habitat):as.factor(Light.Regime)`
d
```

The results are presented below:

	diff	lwr	upr	p adj
Habitat B:Light-Habitat A:Light	-10.40	-11.71065	-9.0893495	2.431388e-14
Habitat A:Dark-Habitat A:Light	-10.15	-11.46065	-8.8393495	2.431388e-14
Habitat B:Dark-Habitat A:Light	-21.25	-22.56065	-19.9393495	2.431388e-14
Habitat A:Light/Dark-Habitat A:Light	-1.00	-2.31065	0.3106505	2.405442e-01
Habitat B:Light/Dark-Habitat A:Light	-11.05	-12.36065	-9.7393495	2.431388e-14
Habitat A:Dark-Habitat B:Light	0.25	-1.06065	1.5606505	9.937246e-01
Habitat B:Dark-Habitat B:Light	-10.85	-12.16065	-9.5393495	2.431388e-14
Habitat A:Light/Dark-Habitat B:Light	9.40	8.08935	10.7106505	2.431388e-14

Habitat B:Light/Dark-Habitat B:Light	-0.65	-1.96065	0.6606505	7.040783e-01
Habitat B:Dark-Habitat A:Dark	-11.10	-12.41065	-9.7893495	2.431388e-14
Habitat A:Light/Dark-Habitat A:Dark	9.15	7.83935	10.4606505	2.431388e-14
Habitat B:Light/Dark-Habitat A:Dark	-0.90	-2.21065	0.4106505	3.541739e-01
Habitat A:Light/Dark-Habitat B:Dark	20.25	18.93935	21.5606505	2.431388e-14
Habitat B:Light/Dark-Habitat B:Dark	10.20	8.88935	11.5106505	2.431388e-14
Habitat B:Light/Dark-Habitat A:Light/Dark	-10.05	-11.36065	-8.7393495	2.431388e-14

From results presented above, we can withdraw the ones which fulfill the null hypothesis, meaning that there is no statistically significant difference between them. We use the function "subset" and the threshold of 0.05:

```
#Subset – Ho=True
no.difference<-subset(d[,c(1,4)],d[,4]>0.05)
no.difference
```

The result below presents the pairs of levels of two analyzed factors for which there is not a statistically significant difference related to seed germination.

	diff	p adj
Habitat A:Light/Dark-Habitat A:Light	-1.00	0.2405442
Habitat A:Dark-Habitat B:Light	0.25	0.9937246
Habitat B:Light/Dark-Habitat B:Light	-0.65	0.7040783
Habitat B:Light/Dark-Habitat A:Dark	-0.90	0.3541739

To continue with the analysis, and plot the results for all combinations of factor levels, we install some additional packages:

```
#Installing packages
install.packages(c("multcompView","emmeans"))

#Loading packages
library(multcompView)
library(emmeans)
```

Now, we can define a linear model by using function "lm," and stating that germination depends on light regime, habitat, and their interaction:

```
#Linear model
model = lm(Germination ~ Light.Regime + Habitat +
Light.Regime:Habitat, data = lab.results.final)
```

In the next step, we calculate the least square means:

```
#Computing the least-squares means
lsmeans.lm = emmeans(model, ~ Light.Regime:Habitat)
```

In previous steps, we performed the analyses of three levels for the factor "light regime" and two levels for the factor "habitat." Therefore, we have six combinations of factor levels that we can adjust by applying the Tukey test once more:

```
#Applying the Tukey test
final<-cld(lsmeans.lm, alpha=0.05, Letters=letters, adjust="tukey")
final
```

Here is the final result:

Light.Regime	Habitat	emmean	SE	df	lower.CL	upper.CL	.group
Dark	Habitat B	60.0	0.32	114	59.1	60.9	a
Light/Dark	Habitat B	70.2	0.32	114	69.3	71.1	b
Light	Habitat B	70.8	0.32	114	70.0	71.7	b
Dark	Habitat A	71.1	0.32	114	70.2	72.0	b
Light/Dark	Habitat A	80.2	0.32	114	79.4	81.1	c
Light	Habitat A	81.2	0.32	114	80.4	82.1	c

Confidence level used: 0.95
Conf-level adjustment: sidak method for 6 estimates
P value adjustment: tukey method for comparing a family of 6 estimates
significance level used: alpha = 0.05

The results include classifying combinations of levels into homogeneous groups, and we can plot them easily:

```
#Plotting the results
p3<-ggplot(final, aes(x = Light.Regime,
y= emmean, fill=.group)) +
  geom_bar(stat="identity")
#Adding labels
  p3<-p3+geom_text(aes(label=.group))

#Additional adjustments
  p3<-p3+facet_wrap(~Habitat)+scale_fill_brewer(palette =
"Set3")+ theme(legend.position = "none")+xlab("Light
regime")+ylab("Least-squares means")
p3
```

Figure 4.6 presents the results.

We can re-arrange Fig. 4.6 and sort the values of germination in ascending order. First, though, we need to make some modifications, and introduce appropriate labels for the combination of levels of the factors we have analyzed. Here is the command:

```
#Adding labels
final$labels<-c("D-B", "L/D-B", "L-B", "D-A", "L/D-A", "L-A")
```

Here, the first letter corresponds to the light regime (D for dark and so on), and the second letter corresponds to the type of habitat (A or B). Now, we can plot the results using these labels:

```
#Plotting the results
p4<-ggplot(data=final, aes(x = reorder(labels,emmean, FUN = max),
y=emmean, fill=.group)) + geom_bar(stat="identity")
#Adding labels
p4<-p4+geom_text(aes(label=emmean), vjust=1)
```

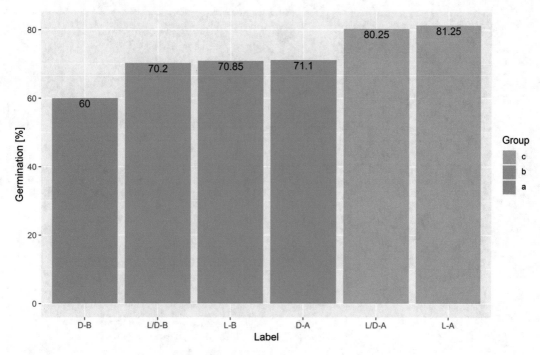

Fig. 4.7 Seed germination [%], different light regimes and types of habitat

```
#Additional adjustments
p4<-p4+ylab("Germination [%]")+xlab("Label")+labs(fill="Group")
p4
```

Figure 4.7 presents the final result.

The results show that seed germination has the lowest value for the combination of levels labeled as D-B which corresponds to 24 h of dark regime and humid habitat, and the highest values of seed germination are observed for the combinations of levels labeled as L/D-A and L-A, which corresponds to 12 h of dark/12 h of light regime and 24 h of light regime, both related to semi-arid habitat, and there is no statistically significant difference between the two of them.

The final Chapter, Advanced R Applications, covers more advanced statistical tests.

Predictive Modeling with Machine Learning Applications

A common motivation of ecological research is to identify the main environmental drivers of ecological processes using empirical modeling. By developing statistical relationships between patterns and processes, empirical models serve two main functions: (1) making accurate predictions over space and time, and (2) making inferences regarding how ecological patterns and processes change across environmental gradients, and determining the relative importance of key variables in driving these patterns.

For example, in a recent study, Shiferaw et al. (2019) evaluated the accuracy of several machine learning algorithms for predicting the geographic distribution and fractional coverage of an invasive shrub, Prosopis juliflora, within the Afar region of Ethiopia. The primary focus of the study was to provide accurate, high resolution species distribution maps to help estimate the socio-ecologic impacts of an invasive species, and for helping develop spatially explicit management strategies to address these impacts. The authors optimized the parameters of each model to reduce model error, compared each model based on their predictive performance, and used the best fitting model to predict the geographic range and fractional coverage of the invasive species.

With the recent wellspring of open-sourced environmental data and the availability of statistical platforms such as R that are capable of carrying out complex machine learning algorithms, few barriers exist to the successful implementation of empirical models. This chapter introduces basic modeling terminology and concepts, and presents a selection of statistical and machine learning models commonly applied to ecological data. Models are then applied to an example dataset for which we identify climatic drivers and limits of western larch (*Larix occidentalis*), a conifer that grows at higher elevations in the western US and Canada. This example application identifies advantages and tradeoffs across models of varying complexity by comparing their prediction accuracy and metrics related to the importance of each predictor variable, as well as how the likelihood of larch presence varies as a function of each predictor variable. Analyses are conducted within the "caret" package (Kuhn 2018), which provides a modular and concise way of manipulating data, training and comparing a wide variety of machine learning algorithms, and using them to make predictions.

© Springer Nature Switzerland AG 2020
M. Lakicevic et al., *Introduction to R for Terrestrial Ecology*,
https://doi.org/10.1007/978-3-030-27603-4_5

5.1   Machine Learning Basics

The type of modeling covered here focuses on what is generally termed "regression modeling," which is a tool used to develop statistical relationships between a set of predictor variables (X) and a single response variable (Y). The response variable (or dependent variable) is the focus of the study, and is something we are trying to predict or better understand how it is related to one or more predictor variables (independent variables, explanatory variables, or features). Many readers will have encountered regression models as far back as high school statistics. Most lessons begin with simple linear regression in which a response variable is regressed on a single predictor variable, and from this model a straight line is drawn through a cloud of points scattered in Cartesian space. Indeed, linear regression is a tool often implemented in a machine learning toolbox. Machine learning and statistics are often described as two different and opposing camps (Breiman 2001a), but much of this debate is outside the scope of the current chapter. For our purposes, machine learning differentiates itself from traditional statistical modeling by stressing efficiency and scalability of models to large data problems, and a desire for complex relationships in the data to be identified and properly modeled by the model (as opposed to the user).

In machine learning parlance, regression modelling is termed *supervised learning* because the data consist of a series of observations that include both a response variable (e.g., relative cover) and several predictor variables (e.g., precipitation, temperature). Alternatively, *unsupervised learning* refers to cases in which there is no response variable, and the goals of analyses are to identify relationships within the data to form groups of like observations or clusters (e.g., hierarchical clustering, k-means, k-nearest neighbors), or to reduce the number of variables in a dataset by transforming the data, and developing new variables (e.g., principal components analysis).

Supervised learning itself comes in two forms depending upon whether the response variable is qualitative (e.g., presence/absence, alive/dead, species), or quantitative (e.g., percent cover, number of trees). Models built for qualitative (or categorical) data are termed classification models, and those for quantitative data, regression models. Regardless of the model type, the goals of supervised learning remain the same: to develop statistical relationships between a set of predictor variables and the response variable for purposes of prediction and inference.

The main output of regression models is a numeric vector of predicted values, while for classification models, main outputs are either a vector of categorical values indicating predicted classes (i.e., presence or absence), or a matrix of probabilities indicating the predicted probability of each class. Another output of importance is a measure of variable importance for each predictor variable. These values are calculated differently depending on the model, but the importance metric helps rank the value of each predictor in modeling the response variable.

Traditional statistical models (e.g., linear regression, logistic regression) often require a few key assumptions regarding the data and their relationships. For instance, linear regression models assume a linear relationship between the response and predictor variables, that variables are independent and approximately normally distributed, that there is little collinearity among predictors, and that errors are normally distributed and homoscedastic (i.e., no significant trend in the model residuals). Interactions among predictors can be included in the model, but have to be explicitly stated in the model. Often, these relationships are not known beforehand and are in fact the motivation behind the modeling in the first place. Machine learning algorithms are generally not free of model assumptions, and in fact assumptions of data independence, and, to a lesser extent, multicollinearity among predictors can still affect machine learning model performance. However, the advantage of these algorithms is that they often do not require *a priori* information regarding the relationships between variables.

The underlying goal of any supervised learning problem is to parameterize the model to minimize some predefined *loss function*. The loss function is a metric that quantifies how well the model fits the

Table 5.1 Common loss functions used in classification and regression problems[a]

Loss function	Formula	Model type
Mean squared error	$$MSE = \frac{1}{n}\sum_{i=1}^{n}\left(Y_i - \hat{Y}_i\right)^2$$	Regression
Mean absolute error	$$MAE = \frac{1}{n}\sum_{i=1}^{n}|Y_i - \hat{Y}_i|$$	Regression
Accuracy	$$ACC = 1 - \left(\frac{1}{n}\sum_{i=1}^{n}|Y_i - \hat{Y}_i|\right)$$	Classification
Log-loss or cross-entropy	$$LL = \frac{1}{n}\sum_{i=1}^{n}Y_i \cdot \log\left(p\left(\hat{Y}_i\right)\right) + \left(1 - Y_i\right) \cdot \log\left(1 - p\left(\hat{Y}_i\right)\right)$$	Classification

[a]Notation is n = number of samples; i is the ith observation in the data. For regression, Y_i is the observed response and $\hat{Y}_i$ is the predicted response for the ith observation. For classification, Y_i is the observed binary response (1 or 0), $\hat{Y}_i$ is the predicted binary response, and $p\left(\hat{Y}_i\right)$ is the predicted probability for the ith observation

dataset. More concretely, the loss function quantifies the amount of error associated with predictions made by the model on the training dataset. It compares model predictions made using a certain parameterization to the actual observed values. Lower values indicate less model error and a better fit to the data. In regression problems, the most commonly used loss function is the mean squared error and for classification, the log-loss error (Table 5.1). The choice of loss function impacts both predictive ability and inference of your final model. For instance, mean squared error tends to give greater weight to large errors as compared to mean absolute error, which may be undesirable if the data contain outliers that can skew results. Similarly, using the accuracy metric for classification may cause problems if large imbalances exist between positive (i.e., presence) and negative (i.e., absence) observations (e.g., 100 positive vs 10,000 negative). A model that fails to predict any positive observations will still have an accuracy of 99.9%, despite its inability to predict the process of interest.

Fitting a model to associated training data requires identifying the optimal set of parameters that minimizes the loss function to the best of the model's ability. Traditional statistical models are often solved using maximum likelihood or matrix algebra, which can cause problems with very large datasets. Machine learning algorithms use an optimization technique known as *gradient descent*, which is used to find the global minimum of a loss function by iteratively updating model parameters and recalculating the loss function until some stopping criterion is met. The basic gradient descent algorithm follows:

1. Initial parameters
   (a) Select a loss function
   (b) Begin fitting the model with random parameters
   (c) Set a learning rate, which determines the size of the steps taken at each step. Lower values (i.e., 0.0001) will take smaller steps, and take many more iterations, while larger values (i.e., 0.1) will take larger steps, but may not converge to the optimal solution.
2. Calculate the gradients of the parameters as the partial derivative of the loss function (i.e., direction and rate of change of the estimated parameter values). The gradient is dependent upon the current parameters, the training data and the selected cost function.
3. Update parameters by adding the results from the gradient multiplied by the learning rate to the current parameter values.
4. Repeat steps 2 and 3 until some criterion is met (i.e., changes in loss function between successive iterations).

Many variants of gradient descent exist that vary by the amount of data used to parameterize the model in each step: batch gradient descent in which all data are used, mini-batch in which a subset of the data is used, and stochastic in which only a single random observation is used.

As an example, we use the "airquality" data from the "datasets" library, which includes daily measurments of air quality for New York City from May to September 1973. We fit a linear model to predict a quantitative response variable, Ozone, to a variety of potential predictor variables including daily solar radiation levels, average wind speed, temperature, and the month and day of the year. The code block below illustrates three methods for estimating model parameters (i.e., coefficients): matrix algebra, maximum-likelihood, and batch gradient descent. This example shows how each method converges on nearly identical coefficient estimates for the linear regression model.

```
library(stats)
library(datasets)

# Preview airquality data
head(airquality)
    Ozone  Solar.R  Wind  Temp  Month  Day
1      41      190   7.4    67      5    1
2      36      118     8    72      5    2
3      12      149  12.6    74      5    3
4      18      313  11.5    62      5    4
5      NA       NA  14.3    56      5    5
6      28       NA  14.9    66      5    6

# Access data, remove NA values and scale the
#   predictor variables
z <- airquality[complete.cases(airquality), ]
x <- as.matrix(subset(z, select = c(Solar.R:Day)))
x.scaled <- scale(x)
y <- z$Ozone

# Using lm function for comparison
mod <- lm(y ~ x.scaled)

#
#
### Matrix Algebra
#
# Need to add column of 1's for the model intercept (design
    matrix)
x <- cbind(rep(1, nrow(x)), x)
x.scaled <- cbind(rep(1, nrow(x.scaled)), x.scaled)

# The symbol %*% is used for matrix multiplication
# solve() takes the inverse of the matrix
mat_algebra <- solve((t(x.scaled) %*% x.scaled)) %*%
t(x.scaled) %*% y

#
#
### Maximum-likelihood estimation
#
```

```r
# Minimize residual sum of squares (i.e., mean squared error)
rss <- function(coef_est, y, x) {
  sum((y - x %*% coef_est)^2)
}

# Initial guesses at parameter estimates
b0 <- rep(0, ncol(x.scaled))

# Minimize rss function using MLE
mle_est <- optim(b0, rss, x = x.scaled, y = y,
                 method = 'BFGS', hessian=TRUE)$par
#
#
### Gradient descent
#
# Squared error loss function
# theta is a vector of coefficient estimates

cost <- function(X, y, theta) {
  m <- length(y)
  betas <- apply(data.frame(x), 1, function(i) i * theta)
  preds <- sapply(data.frame(betas), FUN = sum)
  J <- sum((preds - y)^2) / (2 * m)
  return(J)
}

# theta is a vector of coefficient estimates
gradient <- function(X, y, theta){
  N <- length(y)
  grad <- rep(0, ncol(X))

  for (i in 1:N){
    pred <- sum(theta * X[i,])
    # Derivative of loss function
    grad <- grad + (pred - y[i]) * X[i,]
  }
  return(grad)
}

# Create the gradient descent function
# x is a matrix of predictor values, y is a vector of response
#   values,
# alpha is the "learning rate" that determines the size of the
#   steps taken each iteration
# eta is a cutoff value used to stop the algorithm when
#   additional iterations lead to neglible model improvements
gradient_descent <- function(x, y, alpha = 0.001,
                             num.iterations = 1000,
                             eta = 1e-5) {
  m <- nrow(x)
  num.features <- ncol(x)

  # Initialize the parameters
  theta <- matrix(rep(0, num.features), nrow=1)
```

```
costs <- rep(0, num.iterations)
i <- 0; cost_delta <- 1000

# Update the parameters incrementally
while((i < num.iterations) & (eta < cost_delta)){
  i <- i + 1
  theta <- theta - alpha * gradient(x, y, theta)
  costs[i] <- cost(x, y, theta)
  if(i > 1)
    cost_delta <- abs(costs[i-1] - costs[i])
}

 return(list(theta = theta[1, ], costs = costs[1:i]))
}

# Perform gradient descent to estimate parameters
grad_des <- gradient_descent(x.scaled, y, alpha = 0.005,
                             num.iterations = 1000)

# Collect results from all methods into a dataframe
mod_ests <- data.frame(lm = coefficients(mod),
                       mat_algebra = mat_algebra,
                       mle = mle_est,
                       grad_descent = grad_des$theta)
print(mod_ests)
                lm  mat_algebra      mle  grad_descent
(Intercept)  42.099       42.099   42.099        42.099
   Solar.R    4.583        4.583    4.583         4.585
      Wind  -11.806      -11.806  -11.806       -11.809
      Temp   18.067       18.067   18.067        18.061
     Month   -4.479       -4.479   -4.479        -4.476
       Day    2.385        2.385    2.385         2.384
```

Results confirm that all four methods produce nearly identical parameter estimates. Models intuitatively indicate that ozone concentrations increased with increasing temperature and solar radiation and decreased with increasing windspeed.

A final method for minimizing the cost function of machine learning algorithms covered in this chapter is the *greedy algorithm*, which will be elaborated below in Sect. 5.3.1. This algorithm repeatedly splits the data into smaller and smaller groups by sequentially assigning observations within a group to a smaller subset. Subgroup assignments are determined by the algorithm, which searches all values of all predictor variables to identify a splitting rule that minimizes the cost function across the new subsets. The algorithm is referred to as greedy, because each split is determined locally in isolation, and future steps are not considered.

In this section, we have described the motivation for developing machine learning algorithms, the two types of supervised learning algorithms (regression and classification), and demonstrated how the main goal of these algorithms is to minimize a loss function, to the best of their ability, and to produce predictions with minimal error. This optimization is often done in the background of most machine learning applications in R and not explicitly coded by the analyst. So far in the process, the main choices we must make as the analyst are to determine the type of learning problem we are faced with and the choice of a loss function. These decisions are based on the nature of the dataset, and the questions being asked. The next section discusses and compares some popular machine learning algorithms.

5.2   Model Assessment

In machine learning, model training is the process by which the parameters of a chosen model are iteratively adjusted to minimize the loss function. Model training involves two distinct sets of error assessments: training error, which assesses errors related to predictions made back on the data used to train the model, and validation error, which assesses model error rates on data withheld from model training. Validation error is used to assess how the model generalizes to observations outside the training set, and can inform when model training is complete. Finally, if sufficient data exists, an independent test set of data can be used to quantify the final error rate associated with the fully trained model. The process of evaluating model performance is referred to as *model assessment*.

For regression problems, model assessment metrics are generally related to the amount of variability in the response variable accounted for by the model. That is, how close model predictions are to observed values in reality. Metrics include mean squared error, sum of squared error, mean absolute error, and R-squared. These were covered in the previous section when discussing loss functions (Table 5.1). All metrics assess model error by calculating differences between model predictions and observed values, and summarizing those differences into a single summary metric.

For classification problems, the same basic procedures are followed, but the metrics used to assess model behavior differ from regression. Classification models are assessed in three main ways (1) comparisons of predicted versus observed categorical responses, (2) pseudo-R^2 that approximates the amount of variance explained by the models, and (3) using information theory to develop a receiver-operator curve (ROC), that assesses model performance across a range of threshold settings. For the latter, the term "threshold" refers to the probability cutoff value above which a prediction is considered positive (i.e., predicts the presence of a species). We focus on the first method for model assessment in this chapter. Figure 5.1 shows an example of a confusion matrix and the associated model assessment metrics that can be calculated from this matrix.

A well performing classification model will result in a high proportion of true positive and true negative predictions, which combined will increase the accuracy metric. Sensitivity describes the ability of the model to correctly predict observed positive instances, while specificity focuses on correctly predicting negative observations. Precision is similar to sensitivity, but calculates the proportion of positive predictions that were actually observed positive instances. Kappa and F1 statistics are commonly used as a way of summarizing overall model accuracy for both positive and negative instances.

Machine learning algorithms vary tremendously in their flexibility. Increasing model complexity can help successfully capture statistical relationships within high-dimensional data, but this complexity

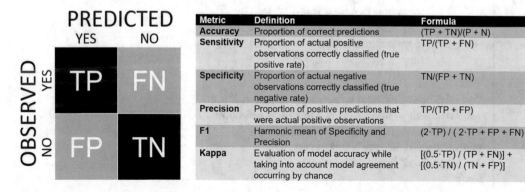

Metric	Definition	Formula
Accuracy	Proportion of correct predictions	(TP + TN)/(P + N)
Sensitivity	Proportion of actual positive observations correctly classified (true positive rate)	TP/(TP + FN)
Specificity	Proportion of actual negative observations correctly classified (true negative rate)	TN/(FP + TN)
Precision	Proportion of positive predictions that were actual positive observations	TP/(TP + FP)
F1	Harmonic mean of Specificity and Precision	(2·TP) / (2·TP + FP + FN)
Kappa	Evaluation of model accuracy while taking into account model agreement occurring by chance	[(0.5·TP) / (TP + FN)] + [(0.5·TN) / (TN + FP)]

Fig. 5.1 Confusion matrix and associated classification model evaluation metrics. Abbreviations are: *TP* true positive, *TN* true negative, *FP* false positive, *FN* false negative

Fig. 5.2 Example of overfitting where training error continues to decrease with increase model complexity, while validation error increases

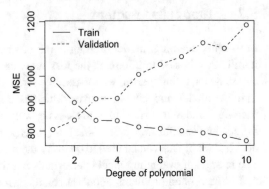

can also come with a cost. Overfitting is a term used to describe the failure of a seemingly well performing model to extrapolate outside of the data used to parameterize the model. Flexibility therefore can lead to an over-tuned model that lacks the ability to generalize outside of the training data. The maxim for this modeling property is known as the bias-variance tradeoff. *Bias* refers to the systematic error associated with models that are not capable of capturing the complexity of the statistical learning problem, regardless of how much data it is given. For example, a linear model that cannot approximate the non-linear relationships among variables will always result in model predictions with high error rates compared to more flexible models (i.e., underfitting). Conversely, *variance* is defined as the amount by which the model would change if it were tuned using a different set of training data.

As an example, we use the "airquality" dataset to fit a series of linear models for which model complexity increases with increasing polynomial degree. Model training error (mean squared error, MSE) was assessed against the validation error. In this example, training error continues to decrease with increasing model complexity while validation error increases (Fig. 5.2), suggesting a high level of overfitting and a failure for complex models to generalize.

```
dat <- airquality[complete.cases(airquality),]
s <- sample(floor(nrow(dat) * 0.7))
train <- dat[s, ]
test <- dat[-s, ]

train_err <- test_err <- c()

for(i in 1:10){
  mod <- lm(Ozone ~ poly(Solar.R + Wind + Temp + Month,
            degree = i), data = train)
  train_err[i] <- mean((predict(mod) - train$Ozone)^2)
  test_err[i] <- mean((predict(mod, newdata = test) -
                  test$Ozone)^2)
}

plot(1:10, train_err, type = 'b', ylim = range(c(train_err,
test_err)), ylab = "MSE", xlab = "Degree of polynomial")
lines(1:10, test_err, type = 'b', lty = 2)
legend('topleft', box.lty = 0, lty = c(1, 2),
    legend = c("Train", 'Validation'))
box()
```

In many circumstances, we lack the ability to adequately assess model error on an independent validation or test set, given constraints on the size of the dataset. In these cases, we turn to resampling methods such as cross-validation for model assessment. The motivation behind resampling methods is to rely on only a single set of training data, but to iteratively train the model on a random sample of the data, and assess validation error rates on the set of data that has been withheld. In this process, each observation is included in both the training and validation data sets, depending on the random sample. The most popular of these resampling methods is cross-validation (CV). For example, leave-one-out CV trains a total of N models, in which each model is trained on all but one withheld observation, and in which N is the total number of observations in the dataset. Model error is then assessed as the mean and deviation of errors across all N test sets. Similarly, k-fold CV first divides the data into k groups and trains the model k times on all but one group, which is used to evaluate model error. K-fold CV can in turn be replicated several times such that group membership varies across replications, which may provide more stable estimates of CV error rates. As an example, we return to the "airquality" data and use the "caret" package to perform 10-fold CV (i.e., $k = 10$). The createFolds function develops a list of length 10, where each list element includes a vector of row numbers to be used for model validation. Linear models are then fit and evaluated 10 times, where the training data for each iteration is a subset of the full dataset that excludes the rows in the respective list element and model error measured on the excluded data.

```
require(caret)
dat <- airquality[complete.cases(airquality),]

k <- caret::createFolds(y = dat$Ozone, k = 10)
cv_err <- sapply(k, function(x){
  train <- dat[-x, ]
  test <- dat[x, ]
  mod <- lm(formula(train), data = train)
  mean((test$Ozone - predict(mod, newdata = test))^2) } )
print(c(mean(cv_err), sd(cv_err)))

[1] 454.4019 316.7507
```

5.3   Machine Learning Algorithms

Machine learning algorithms run the gamut of complexity from linear and generalized linear models to artificial neural networks and beyond. We have already used linear models in an example application in the previous section. Linear models are a very efficient and effective model on occasions where Y is linearly related with the matrix of predictor variables. When this constraint is not satisfied by the data, alternative machine learning algorithms are capable of modeling more complex relationships with similar efficiency. We describe four popular machine learning algorithms of varying complexity, and use them in a subsequent analysis to compare their performance. Model complexity itself is difficult to quantify across models as they lack true parameters in the sense of traditional statistical models. Rather, they incorporate weights and rulesets used to partition the data in multidimensional space. Models in this section are presented in order of increasing complexity relative to one another.

5.3.1   Decision Trees

Tree-based methods attempt to partition the data into smaller and smaller groups, in which group membership is determined by a series of binary splits on the predictor variables (De'ath and Fabricius 2000, Breiman 2017). Starting with the full set of training data, the algorithm searches all possible values of all predictor variables and each data point is put into one of two possible groups. The first group includes all data with $s \leq X_j$ and the second group includes all data with $s > X_j$, where s is the value of the split for predictor variable X_j. The algorithm then calculates the mean within-group variability to determine the value of j and s that minimizes this variation. Common measures of variation include mean square error for regression and the Gini index for classification problems.

This process is then repeated for each group until some stopping criterion is reached. As we will see, the resulting architecture of the model can be visualized as a tree, with each split referred to as a *node*. Predictions are made by fully trained trees to new data by comparing the predictor values of each new observation to each of the binary splits beginning with the *root node* until it reaches a *terminal (or leaf) node* where no more splits are made. For regression, the predicted value for the new observation is the mean of the training observations in the terminal node; for classification, it is the majority class. As the tree grows, the number of training observations within each terminal node decreases and predictions are based on fewer and fewer training data. Furthermore, with each data partition, the loss function is reduced until, theoretically, the number of terminal nodes equals the number of training observations (i.e., 0 error). This is an example of the greedy algorithm mentioned in the previous Sect. 5.1. Consequently, model complexity increases with the number of data partitions and large trees often suffer from overfitting.

Overfitting can be curbed by adjusting a *complexity parameter*, which will only add an additional split of the data, if the split will improve the overall model fit by some specified amount. Larger complexity parameters generally result in smaller trees (fewer splits), given the bigger the hurdle for model improvement at each split. This parameter can be determined through cross-validation, which balances tree size with predictive ability.

Returning to the "airquality" dataset we fit a regression tree model to the data using the "rpart" function in the eponymous package. We first fit a model with a very low complexity parameter to build a large tree. We then compare CV error rates across a range of complexity parameters to identify the model with the fewest splits of the data so that additional splits lead to minor improvements in model accuracy.

```
require(rpart)
## Loading required package: rpart
dat <- airquality[complete.cases(airquality), ]
r <- rpart(formula(dat), data = dat,
          control = rpart.control(cp = 0.001))

plotcp(x = r, cex.axis = 0.7, cex.lab = 0.8)
```

We see (Fig. 5.3) that the CV error decreases as the complexity parameter (cp) decreases (smaller hurdle for adding additional splits, and larger trees). The CV error rates on the y-axis are relativized by dividing CV error sum of squares by the total sum of squares for the data. Tree size, on the top axis, is the number of leaf nodes. The horizontal line represents the mean + standard CV error of the model with the lowest CV error. This can be used to select the model with the largest complexity parameter value below the horizontal line.

Fig. 5.3 Plot of the relationship between cross-validation error and the decision tree complexity parameter. The number of splits in the decision tree increases from left to right (size of tree) as the complexity parameter (cp) decreases. The horizontal line represents the minimum average cross-validation error plus the standard deviation of error. The point with the largest cp value below the horizontal line is the recommended parameter value

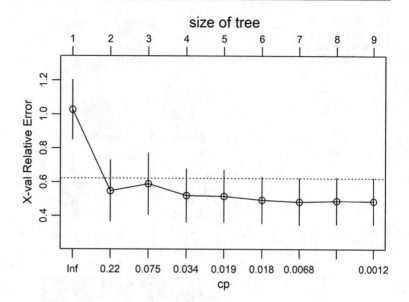

Next, we compare two regression trees with varying complexity parameters.

```
require(rpart.plot)
# Lower complexity parameter, larger tree
rp_big <- prune(r, cp = 0.018)
# Larger complexity parameter, smaller tree
rp_small <- prune(r, cp = 0.075)

# Set up graphing space to plot two rpart trees side-by-side
par(xpd = NA, mfrow = c(1, 2), omi = c(0.1, 0.05, 0.05, 0.05),
    mai = c(0, 0.15, 0, 0.15))
rpart.plot(rp_big)
text(x = 0, y = 1, labels = 'A')
rpart.plot(rp_small, tweak = 0.6)
text(x = 0, y = 0.96, labels = 'B')
```

The tree to the left (cp = 0.018) has 6 terminal nodes on 5 splits of the data, while the tree to the right (cp = 0.075) has 3 terminal nodes on 2 splits of the data (Fig. 5.4). Splitting rules are given at each node, and determine the direction of data down the tree, starting from the root node at the top of the tree. The specific directive at each node (i.e., Temp < 83) directs the data that satisfies the rule to the **left**. So, in both trees we see that predictor variable "temperature" is used as the first partition of the data at the root node. All observations with temperatures <83 proceed down the left of the tree and are further split by the predictor "wind." Observations with temperatures ≥83 proceed to the right of the tree and into a terminal node for the model on the right. These observations would be predicted to have an ozone level of 77 ppb. The larger tree on the left further splits observations in this node by wind speed and again by temperature. Results from these models suggest that high temperatures and low wind speeds lead to higher ozone levels.

We can see this further using the "plotmo" package to plot the partial dependence plots for the model with two splits (right column). These plots (Fig. 5.5) show the ozone response modeled over the range of the temperature and wind variables. For partial dependence plots of a given predictor at each point the effect of the background variables is averaged.

Fig. 5.4 Example regression trees of varying complexity (**a**) cp = 0.018, and (**b**) cp = 0.075

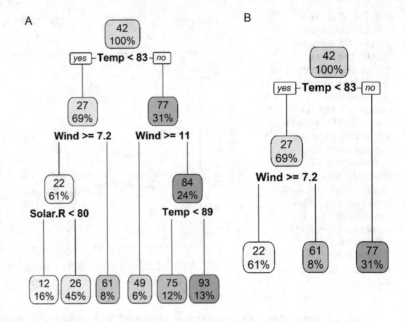

Fig. 5.5 Example partial plots resulting from a decision tree model relating air quality (ozone concentration, ppb) to wind speed (mph) and temperature (F). Plots depict the marginal effect of each predictor on the response variable

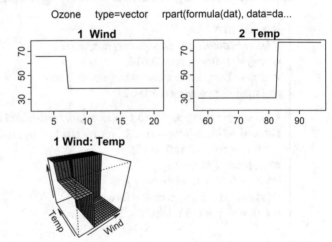

```
require(plotmo)
plotmo(rp_small, pmethod="partdep")
```

These plots clearly show the distinct step-like function that regression trees produce, which depict the binary splits at each tree node. We can further see how the model partitions the data space in the two-dimensional plot below (Fig. 5.6). Plotting the original data back onto the partial plot further verifies high correspondence between model predictions and observed ozone concentrations.

```
require(lattice)
require(pdp)

p <- pdp::partial(rp_small, pred.var = c('Temp', "Wind"))
pdp::plotPartial(p)
lattice::trellis.focus()
```

Fig. 5.6 Two-dimensional decision space representing decision tree predictions of air quality in relation to wind and temperature predictor variables

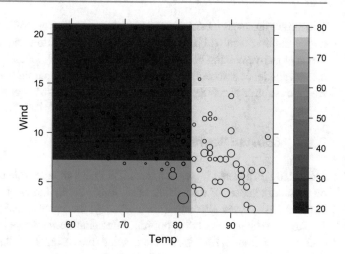

```
# Range the size of the points (cex) between 0.2 and 2 based on
#   their measured ozone concentration
pt_size <- (dat$Ozone - min(dat$Ozone)) * (2 - 0.2) /
           (max(dat$Ozone) - min(dat$Ozone)) + 0.2

# Plot airquality data on top of the partial plot
panel.points(x = dat[, c('Temp', 'Wind')], cex = pt_size,
           col = 'black')
```

Unfortunately, while decision trees are easy to implement and interpret, they generally have poor predictive ability and high variance compared to other regression and classification techniques. However, as we shall see in subsequent sections, predictive performance can improve considerably when many decision trees are combined using ensemble modeling techniques.

5.3.2   Random Forest

Random forest (RF) is an ensemble modeling technique that uses ***bootstrap aggr****egation,* or *bagging* to combine many individual decision trees each trained on a random subset of the training dataset and calculates average predictions across the trained models (Breiman 2001b, Cutler et al. 2007, James et al. 2013). By randomly sampling the training data across decision trees, bagging effectively reduces the high variance associated with single decision trees that are sensitive to the specifics of the training data, and do not readily generalize beyond these data. RF improves upon bagging by also selecting not only a subset of the training data for each decision tree, but also randomly samples the predictor variables. This further increases the *independence* of each decision tree, which in turn reduces model variance and improves predictive accuracy. Trees within RF are generally left unpruned and can therefore be quite large, but overfitting is reduced by averaging predictions across many independent trees. As we will see in the example below, RF, along with other ensemble models, are capable of modeling complex interactions and highly non-linear relationships.

Parameters, which are generally tuned to improve RF performance for the specific application, are the proportion of predictor variables selected during each model iteration (*mtry* parameter), and the total number of trees used to build the RF model (*ntree*).

Besides improved predictive capability, other additional attributes that make RF an appealing model for ecological research are the use of out-of-bag model error statistics and variable importance measures.

The former produces model error estimates similar to cross-validation, in which model error is assessed on the training data withheld during each model iteration and across all iterations. The latter allows for an objective ranking of the relative influence of each variable on model performance. Furthermore, the independence associated with each tree lends RF to parallel programming, which can greatly reduce the time required to run the algorithm, which is particularly helpful for big data problems.

5.3.3   Boosted Regression Trees

Similar to RF, boosted regression trees (BRT) are an ensemble of decision trees, and each tree is built on a random subset of data (Elith et al. 2006, 2008, James et al. 2013, Elith and Leathwick 2017). However, BRT uses alternative methodology for training and combining individual decision trees. Unlike RF, which grows large, unpruned decision trees using a subset of the predictor variables during each model iteration, BRT grows many small trees using the full set of predictors. Boosting is a sequential stage-wise algorithm, in which individual decision trees are built sequentially, and each tree is dependent on the predictions made by previously grown trees. The first tree in the model is trained using the usual methods outlined in Sect. 5.3.1. Decision trees, but the response variables for all subsequent trees are the residuals of the model built up to that point. Furthermore, as iterations progress, the boosting algorithm increases the weight of training observations that are poorly predicted by the existing set of trees, with the goal of reducing the loss function by emphasizing high residuals. In contrast to RF, BRT exploits the *dependence* among decision trees. By relying on many small, weak learners, the algorithm learns slowly. An additional *shrinkage parameter* further slows down the algorithm by reducing the level of influence each tree has on the final model, similar to the alpha parameter used in gradient descent in Sect. 5.1. Machine learning basics. Combined, boosting reduces both bias (through the stage-wise progression of the model) and variance (by averaging across models).

Model parameters that require user input are (1) the number of splits for each individual decision tree (*interaction depth,* usually between 1 and 3), (2) the number of trees in the BRT model, and (3) the shrinkage parameter value (usually between 0.01 and 0.001). Given the sequential nature of the algorithm, and the focus on training observations with high residuals, increasing the number of trees beyond a point can lead to overfitting. This can be addressed using cross-validation to identify the point at which increasing the number of trees leads to higher CV error.

Model development can be slower for BRT compared to RF, but parallel version implementations, such as XGboost, are becoming more popular, particularly for big data problems.

5.3.4   Stacked Ensemble

Model stacking is another ensemble learning technique that combines multiple fully trained models (level-one base learners) by using predictions made by these models as predictor variables in a second-level metamodel (Wolpert 1992, Breiman 1996, Friedman et al. 2001). Unlike RF and BRT, stacking aims to create ensemble predictions from a set of strong and diverse base models. The algorithm proceeds by training each of the level-one base learners on the full set of training data, and once tuned, collects predictions made by each base learner using k-fold CV. These predictions then form a new matrix of predictor variables along with the original response variable. The metamodel is then trained on the newly created data, which can then be used to make predictions to a test set. The analyst must choose the number and type of base learners, and the type of metamodel. The metamodel itself can be any number of models; in the example below, two stacked models are developed, one using a logistic regression metamodel and another using a random forest metamodel.

5.4   Case Study: Modeling the Distribution of Western Larch (*Larix occidentalis*) Under Recent Past and Future Climate Across Western North America

We present a machine learning workflow for developing a model to predict the geographic distribution of western larch (*Larix occidentalis*) across western North America under various climate scenarios. This example highlights some of the important aspects of the modeling process including (1) developing hypotheses of the relationships between the processes and associated environmental drivers of interest, (2) data gathering and feature reduction, (3) model evaluation and selection, and (4) developing and mapping model predictions.

5.4.1   Generating Hypotheses

A common misconception of machine learning is that, given the "black box" nature of the algorithms and their ability to efficiently model complex relationships with little user input, it is therefore appropriate to indiscriminately include as much data and as many predictors as possible for the algorithm to sort out and produce a viable model. Given the enormous amount of freely available data, it is tempting to gather data and begin modeling with only a vague idea of the proposed research and no plan to evaluate model outputs. Charting this course will inevitably lead to models that underperform or lack the ability to make appropriate inferences regarding the modeled relationships.

The crucial first steps of model development are to (1) understand the study system through a review of past research conducted on the topic of interest, (2) develop testable research questions and hypotheses, (3) identify data requirements, (4) determine how model outcomes will be evaluated and how they will address the research questions, and (5) anticipate how the model results might advance previous work and contribute to the current state of the science.

For our example, we are interested in developing a model to better understand the main climatic determinants of the current geographic distribution of western larch, and to predict where a suitable climate space for this species will exist under future climate change. Western larch is a deciduous conifer species whose range extends into the mountainous regions of the eastern Cascades in Washington and Oregon and the Rocky Mountains and foothills of southeastern British Columbia, northern Idaho and northwestern Montana. It is the largest larch species in North America, and mature western larches are the most fire tolerant tree species in its range. Combined with its high value as a timber species, as well as the important habitat it provides for a variety of wildlife species, western larch is an important and unique component of western conifer forests. Western larch is generally found at mid-elevations and within a fairly restricted geographic range. Past research shows that this distribution is largely driven by climate, with warm temperatures and low precipitation limiting at its southern extent and lower elevations, and cold temperatures to the north and at upper elevations. Given the purported climatic controls on western larch, we developed five main research questions outlined in Table 5.2.

5.4.2   Data Gathering and Preparation

We used a spatial dataset from Little Jr. (2006) to capture the current geographical distribution of western larch across North America[1]. The file was downloaded to a local drive and accessed in R using "rgdal" package. The following code accesses the spatial polygon shapefile representing the current

---

[1] https://databasin.org/galleries/5e449f3c91304f498a96299a9d5460a3#expand=12211

Table 5.2 Research questions for the development of machine learning algorithms used to model the distribution of western larch

Question	Premise	Hypothesis	Test
How well do climate variables predict the geographic range of western larch?	The range of western larch is known to be limited by climate, but other factors may also be limiting and there omission from the model may result in low model accuracy	A climate-only model will accurately predict the current distribution of western larch	Assess model accuracy on an independent test set of data
Are models of high complexity required to adequately develop statistical relationships between climate predictor variables and western larch presence/absence?	Species are distributed across complex environmental gradients that often require more complex models to capture these relationships	Ensemble models capable of modeling complex relationships will perform better than simpler models	Compare model accuracy across models ranging in complexity
Which climate variables are most influential to limiting the distribution of western larch?	A combination of climate factors are thought to limit larch, but these variables are most important? Is precipitation more important than temperature? Are extreme conditions more important than averages? Possible climate change implications	Seasonal variables will have higher importance than annual climates	Compare variable importance for seasonal and annual climate variables
How does the predicted probability of larch presence change across the range of key climate drivers?	Visualizing the direction and shape of the response curve, which depicts changes in the response variable with increasing/decreasing values of a predictor variable reveals the functional form of their relationship and possibly can identify threshold behavior	Models will identify non-linear relationships for climate predictor variables, suggesting abrupt changes in western larch tolerances along steep climate gradients in the region	Partial dependence plots relating the marginal effect of a predictor variable on the probability of larch presence
How does the climate niche predicted by our model shift under future climate change?	The climate niche space occupied by a species indicates their tolerances to precipitation and temperature regimes, outside of which the species cannot grow and reproduce. Under climate change these joint conditions will not likely persist in their current geographic distribution but will shift across the landscape, suggesting a possible future range shift for the focal species	The climatic envelope for western larch will shift north and to higher elevations compared to its current distribution	Use the trained machine learning algorithm to predict the probability of western larch presence under predicted future climate conditions

geographic distribution of western larch and projects it into latitude-longitude coordinates. A shapefile is a spatial dataset commonly used in geographic information systems (GIS) for mapping and analysis. From there, we can get the geographic coordinates from the shapefile so we can map the geographic data using the "ggmap" library, which uses similar methods and nomenclature as the "ggplot2" library used in Chaps. 2, 3, and 4 (Fig. 5.7).

```
require(rgdal)
require(raster)
require(ggplot2)
require(ggmap)
require(plyr)
require(ggpolypath) # required to plot SpatialPolygons with holes
```

Fig. 5.7 Proximity map created using the "ggmap" package depicting the geographic distribution of western larch. Basemap depicts forest vegetation in deep green and nonforest vegetation in light green, blue indicates water and grey shows alpine environments. Western larch distribution is overlain in dark grey

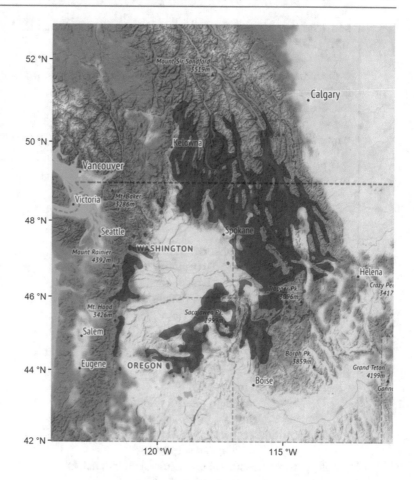

```
# Set local directory
inDir <- "D:/Dropbox/Book_R_EnvScie"

# Tree species geographic distribution (shapefile)
# Little Jr., E. 2006. Digital representations of tree species
#    range maps from 'Atlas of United States Trees'.
# Downloaded from:
#https://databasin.org/galleries/5e449f3c91304f498a96299a9d5460a3#exp
and=12211
spd_merc <- readOGR(paste0(inDir,
'/Data/species_range_maps/Little_range_maps/WL/lariocci.shp'))

# Subset data to remove "holes" in the distribution of western
#    larch within the polygon perimeter
spd <- subset(spd_merc, subset = CODE == 1)

# Project shapefile to latitude-longitude
spd_ll <- spTransform(spd, CRSobj = CRS("+init=epsg:4326"))

# Create a dataframe of the polygon coordinates for mapping
spd_ll.f <- fortify(spd_ll)
# Regions defined for each Polygons
```

```
spd_ll$id <- row.names(spd_ll)  # provide same column names for join
spd_ll.f  <- join(spd_ll.f, spd_ll@data)

# Grab the extent of the distribution of western larch and make it
#   bigger
ext <- extent(spd_ll) * 1.5
ext <- unlist(head(attributes(ext), -1))
names(ext) <- c('left', 'right', 'bottom', 'top')

# Grab a basemap to for geographic context
m <- get_map(location = ext)

# Create custom x- and y-axis labels
ewbrks <- seq(-125, -110, 5)
nsbrks <- seq(42, 56, 2)
ewlbls <- unlist(lapply(ewbrks, function(x)
                ifelse(x > 0, paste(x, "°E"),
                ifelse(x < 0, paste(abs(x), "°W"), x))))
nslbls <- unlist(lapply(nsbrks, function(x)
                ifelse(x < 0, paste(abs(x), "°S"),
                ifelse(x > 0, paste(x, "°N"), x))))

# Basemap and western larch distribution
ggmap(m, extent = 'panel', padding = 0) +
  geom_polypath(data = spd_ll.f,
                  mapping = aes(x = long, y = lat,
                  group = group, alpha = 0.4)) +
  scale_x_continuous(breaks = ewbrks, labels = ewlbls,
                  expand = c(0, 0)) +
  scale_y_continuous(breaks = nsbrks, labels = nslbls,
                  expand = c(0, 0)) +
  guides(fill = F) +
  theme(legend.position = "none") +
  xlab("") +
  ylab("") +
  theme(plot.margin=grid::unit(c(0,0,0,0), "mm"),
          axis.text = element_text(size = 8))
```

A digital elevation model was used to characterize elevation gradients across North America:

```
# DEM from:
# https://databasin.org/datasets/d2198be9d2264de19cb93fe6a380b69c
# ~1km resolution (927 m)
dem <- raster(paste0(inDir, "/Data/species_range_maps/elevation/NA/na_dem.tif"))
```

Finally, we downloaded historical and future projected climate data from Hamann et al. (2013)[2]. These data covered all of western North America west of 100° longitude, and are comprised of 11 million grid cells at 1km resolution, and designed to capture long-term (30 years) climate gradients,

---

[2]Downloaded from: https://sites.ualberta.ca/~ahamann/data/climatewna.html

temperature inversions, and rain shadows. The historical climate data are average conditions for the years 1961–1990. Future climate projections represent predicted conditions for the years 2071–2100, based on the Coupled Model Intercomparison Project phase 3 (CMIP3) of the World Climate Research Programme (Meehl et al. 2007). Future climate variables were simulated using the Ensemble of 23 Atmosphere-Ocean General Circulation Models run under the A2 emissions scenario, representing "worst case" emissions. These models are used in the International Panel on Climate Change and elsewhere. More information can be found in Brekke et al. (2013). See Appendix 3 for a description of the climate predictor variables used in the analysis.

The climate rasters were combined into a RasterStack object, which is essentially a list of rasters, but stacking ensures that all rasters in the stack share the same projection, extent, and resolution. The RasterStack was then sampled at 10 km spacing to develop a data frame of predictor values with one row of climate data for each sample point. We opted for a regular spacing of sample points rather than a random sampling to reduce the level of spatial dependence among the sample points. Spatial dependence arises from a process known as spatial autocorrelation, which describes the property of spatial systems, in which things close to one another tend to be more alike than those further away (Tobler's first law of geography, (Miller 2004)). Removing spatial dependency in the sample helps ensure independence among observations in the sample, which is an essential property in statistical modeling. This resulted in a sample size of 108,077.

```r
# Climate data (1km rasters), representing 24 biologically
#   relevant variables, including seasonal and annual means,
#   extremes, growing and chilling degree days,
#   snow fall, potential evapotranspiration, and a number of
#   drought indices
# Data from:
#http://www.cacpd.org.s3.amazonaws.com/climate_normals/NORM_6190_Bioclim_
ASCII.zip
clim <- stack(list.files(paste0(inDir,
'/Data/climate/NORM_6190_Bioclim_ASCII'), full.names = T, pattern = "\\.asc$"))

# Project DEM raster to the same projection as the climate data
dem <- projectRaster(dem, to = clim[[1]])

# Add elevation model to the stack (27 layers)
clim$elev <- dem
# Simplify naming convention
nms <- unlist(lapply(strsplit(names(clim), '_'),
               function(x) x[3]))
# Rename some of the longer layer names.
nms[c(19, 20, 22, 23, 27)] <- c('PPT_sm', 'PPT_wt', 'Tave_sm',
                                 'Tave_wt', 'Elev')
names(clim) <- nms

# Need to sample the landscape to reduce the number of total data
#   points such that sample points are spaced out to reduce
#   spatial dependency among points
# Generate a sample of the climate and elevation data such that
#   sampled points are 10,000 meters apart
e <- extent(clim)

s <- expand.grid(x = seq((e@xmin + 500), (e@xmax - 500),
```

```
                                by = 10000),
                        y = seq((e@ymin + 500), (e@ymax - 500),
                                by = 10000))

# Based on the above coordinates, grab the cell ID numbers for the
#   climate rasters
cn <- cellFromXY(clim, xy = s)
s$cellNum <- cn

# Create a SpatialPointsDataFrame from the coordinates, and assign
#   the appropriate projection
coordinates(s) <- ~x + y
proj4string(s) <- proj4string(clim)

# Climate boundary downloaded from same site as the climate data
# Downloaded from:
#http://www.cacpd.org.s3.amazonaws.com/climate_normals/Reference_files_ASCII.zip
bounds <- readOGR(paste0(inDir,
                        '/Data/climate/Elevation/Bound_WNA.shp'))

# Extract climate and elevation data from the subset points
x <- clim[cn]
w <- which(complete.cases(x))
x <- data.frame(x[w, ])
s_sub <- s[w, ]

# Create binary response variable: yes if the sample point is
#   within the boundaries of the western larch distribution, and
#   no, if not
spd_lam <- spTransform(spd, CRSobj = CRS(proj4string(clim)))
s_tree <- s_sub[spd_lam, ]
cn_tree <- cellFromXY(clim, xy = coordinates(s_tree))

x$Presence <- factor(ifelse(s_sub$cellNum %in% cn_tree,
                        'yes', 'no'), levels = c('yes', 'no'))
x$x <- coordinates(s_sub)[, 1]
x$y <- coordinates(s_sub)[, 2]

x <- subset(x, select = c(Presence, AHM:Elev, x, y))
coordinates(x) <- ~x + y

# Remove excess points that do not have an elevation value
# Final data set for model training including climate normals
#   (1961 -1990) and elevation
x <- subset(x, subset = !is.na(Elev))

# Predicted future climate (2071 - 2100) using the Ensemble CMIP3
#   and A2 emissions scenario downloaded from:
#http://www.cacpd.org.s3.amazonaws.com/ensemble_AOGCMs/ENSEMBLE_A2_2080s_
Bioclim_ASCII.zip
clim_A2 <- stack(list.files(paste0(inDir, '/Data/climate/
                        ENSEMBLE_A2_2080s_ASCII'),
                        full.names = T, pattern = "\\.asc$"))
```

5.4.3   Model Development

We modeled the distribution of western larch presence as a function of climate and elevation using a suite of machine learning algorithms (library) including:

1. Logistic regression (stats)
2. Classification trees (rpart)
3. Boosted regression trees (gbm)
4. Random forest (ranger)
5. Stacked ensemble of models 1–4 using a logistic regression meta-model (caret)
6. Stacked ensemble of models 1–4 using a RF meta-model (caret)

We began by splitting the data into separate training and testing sets, given the large amount of data. The correlation matrix of the predictors was assessed to identify multicollinearity among the variables.

```
require(caret)
require(caretEnsemble)
require(ggcorrplot)
require(ranger) # Random forest implementation
require(gbm) # Boosted regression trees
require(rpart) # Decision trees

# Grab the data frame from the SpatialPointsDataFrame created
#   above
z <- x@data

# Calculate correlation matrix across just the predictor variables
cormat <- round(cor(z[, -1]), 2)
g <- ggcorrplot(cormat, hc.order = TRUE, method = "circle")
print(g)
```

The correlation plot (Fig. 5.8) shows a high degree of multicollinearity among predictor variables, suggesting that a redundancy of information among the variables. Reducing the number of predictors will make the models more interpretable. Linear and logistic models multicollinearity reduces the accuracy of coefficient estimates, increases their associated standard deviations, and reduces overall model power to detect significant coefficients. This effect is less apparent in decision trees and the ensemble methods, but it still effects variable importance metrics and model interpretability.

For these reasons, highly correlated variables were eliminated from analyses. The "findCorrelation" function in the "caret" package searches through the correlation matrix, and, if two variables have a high correlation, the function looks at the mean absolute correlation of each variable and removes the variable with the largest value.

```
# Find correlated variables and remove them
rm_vars <- findCorrelation(x = cor(z[, -1]), cutoff = 0.7,
                           names = F)
x_vars <- names(z[, -1])[-rm_vars]
```

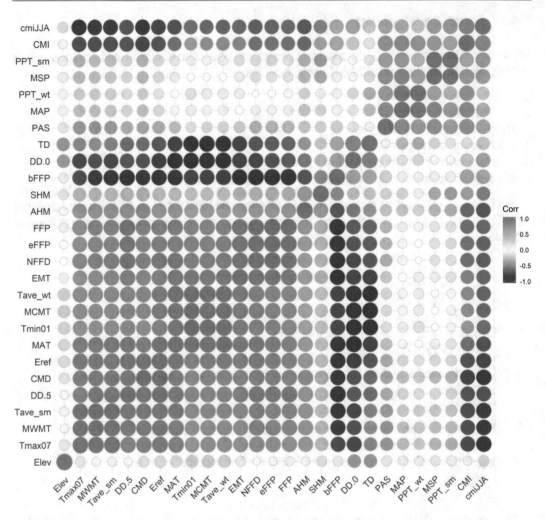

Fig. 5.8 Visualization of the correlation matrix using the library ggcorrplot for the elevation and climate normals data used as predictor variables in the western larch species distribution modeling

This reduces the number of potential predictors from 27 to 7. The level of collinearity is not surprising, becasue many of the climate variables are derived from other climate variables (i.e., summer, winter, and annual precipitation are all derived from monthly precipitation values). Given the small number of predictors that remain in the model, there is no need to use any *feature reduction* methods to further eliminate potential predictor variables. Methods such as forward and backward elimination are often done to retain only the most important predictors in an effort to balance model parsimony with model performance—that is, model interpretability with model accuracy (Kuhn 2012). However, several methods exist for feature reduction; for instance, in the "caret" package, the function "rfe" performs backwards elimination so that the variable with the lowest variable importance is removed from the model iteratively, until a single predictor remains. During each iteration, model accuracy is assessed using cross-validation.

The final set of seven predictor variables are shown in Table 5.3 and Fig. 5.9. These predictors represent a variety of seasonal precipitation, temperature, and drought conditions.

The full set of sample points was then split into separate training (70% of data) and testing (30%) sets.

Table 5.3 Final set of predictor variables[a]

Variable	Definition	Units
AHM	Annual heat-moisture index	(°C 10) m^{-1}
bFFP	Beginning date of frost free period	Julian day
DD.0	Number of degree days below 0°C	°C
Elev	Elevation	m
PPT_sm	Summer precipitation	mm
PPT_wt	Winter precipitation	mm
SHM	Summer heat-moisture index	(°C 10) m^{-1}

[a]All climate variables depict 30-year climate normal conditions for the years 1961–1990 estimated using the climateWNA algorithm.

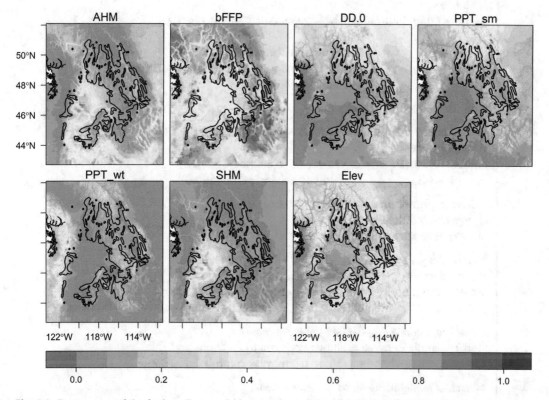

Fig. 5.9 Raster maps of the final predictor variables used in modeling the geographic distribution for western larch (black outlines) in western North America (see Table 5.3 for variable definitions) using the rasterVis package. Variables were scaled between 0 and 1 using a linear transformation of the native scale for each variable within the region shown

```
# Partition the data into training (70%) and testing (30%) sets
s <- createDataPartition(y = z$Presence, times = 1,
                p = 0.7, list = F)
train <- z[s, ]
test <- z[-s, ]
train <- subset(train, select = c('Presence', x_vars))
test <- subset(test, select = c('Presence', x_vars))
```

The "caretList" function is used to run the logistic regression, classification tree, GBM, and RF models, and is a wrapper for the main workhorse function "train" within the "caret" package. The "train" function itself uses a grid search of values for key variables for each model. The "trainControl" function tells the algorithms to assess model fit under each unique set of parameters, using10-fold CV repeated three times. The "train" function then assesses model CV error for models tuned across all combinations of parameters, using the default Accuracy error metric.

```
require(doParallel)

# Establish control parameters for caret models for training
#   models and selecting which elements to save off
# See: https://rpubs.com/zxs107020/370699
# https://cran.r-project.org/web/packages/caretEnsemble/vignettes/caretEnsemble-
intro.html
control <- trainControl(method = "repeatedcv", number = 10,
                        repeats = 3,
                        createResample(train$Presence, 10),
                        savePredictions = TRUE, classProbs = TRUE,
                        verboseIter = TRUE)

mod_list <- c("glm", 'rpart2', 'gbm')

# Need to run this for random forest to calculate variable
#   importance
tuneList <- list(ranger = caretModelSpec(method="ranger",
                            importance = "permutation"))

# Start multicore processors using four cores
cl <- makePSOCKcluster(4)
registerDoParallel(cl)

# logistic regression, CART, BRT, and RF models run using
#   caretList
mods <- caretList(Presence ~ ., data = train, trControl = control,
                  methodList = mod_list, tuneList = tuneList, metric = "Accuracy")

# Stacked model using GLM as the meta-model
mod_ensemble <- caretEnsemble(mods, metric = "Accuracy",
                              trControl = control)

# Stacked model using RF as the meta-model
mod_stack <- caretStack(mods, method = "rf",
                        metric = "Accuracy", trControl = control)

stopCluster(cl)

# Combine models in the caretStack, and re-order them
mods_all <- mods
mods_all$glm_stack <- mod_ensemble
mods_all$rf_stack <- mod_stack
mods_all <- mods_all[c(2, 3, 4, 1, 5, 6)]
```

Once the series of models are tuned, the test data are used to assess final error rates. The "confusionMatrix" function compares the observed presence/absence data from the test set to the predictions made by the individual models on the test set of predictor variables. Predictions are made assuming no knowledge of the actual response when making predictions. The "lapply" function is used to call functions, such as "predict", for each of the models in the caretList of trained models in lieu of using a "for" loop over each element of the list.

```
p <- lapply(mods_all, predict, newdata = test, type = 'prob')
names(p) <- names(mods_all)
p <- lapply(p, function(i) if(class(i) == 'numeric') i else i[, 1])
p <- lapply(p, function(i) as.factor(ifelse(i > 0.5, 'yes', 'no')))
names(p) <- names(mods_all)

mod_accuracy <- lapply(p, function(x)
                    confusionMatrix(reference = test$Presence,
                        data = x, positive = 'yes'))
mod_accuracy <- lapply(mod_accuracy, function(i)
                    c(i$overall[1:2], i$byClass[7],
                        i$byClass[c(1, 2, 5)]))
mod_accuracy <- data.frame(do.call(rbind, mod_accuracy))
print(mod_accuracy)
```

5.4.4   Model Assessment

Some interesting model error results are shown in Table 5.4. Few differences in accuracy (first column) are apparent across models. However, the sensitivity and precision metrics suggest that models varied in their ability to correctly predict the presence of western larch. This is likely caused by the imbalanced design of our model, in which the number of absences outweigh the number of presences 65:1. The problem with using the accuracy metric for evaluating the performance of classification models under imbalanced designs is illustrated by the GLM result in the first row of Table 5.4. This model predicted no western larch presences, yet returned an accuracy of 0.984. Across all metrics, the RF model and the stacked RF meta-model performed the best.

The RF model was chosen for the final model used to predict western larch range under previous (1961–1990) and future (2080s) climate (Fig. 5.10).

Differences in model accuracy can be illustrated through mapping model predictions with respect the current range of western larch (Fig. 5.10). These maps depict the predicted probability of western larch presence, where predicted probabilities >0.5 (red colors) suggested larch presence. Overall, the spatial accuracy of the models was similarly good, with the exception of logistic regression (GLM). The decision tree model (CART) and GBM both correctly predicted about half of the 1 km cells within the range of western larch, while RF and the stacked models correctly predicted close to two-thirds of the cells (Fig. 5.11). Most models predicted western larch occurrence outside and to the west of the current range of western larch, suggesting that climate conditions may be favorable for western larch in this area, and some other limiting factor prevents its occurrence.

Table 5.4 Model performance metrics based on model predictions to the test set

Model	Accuracy	Kappa	F1	Sensitivity	Specificity	Precision
GLM	0.984	−0.003	NaN	0.000	0.998	0.000
CART	0.989	0.526	0.532	0.431	0.997	0.694
GBM	0.989	0.547	0.553	0.466	0.997	0.679
RF	*0.991*	*0.641*	*0.646*	*0.571*	*0.997*	*0.743*
GLM stack	0.988	0.375	0.380	0.245	0.999	0.838
RF stack	0.987	0.580	0.587	0.635	0.992	0.545

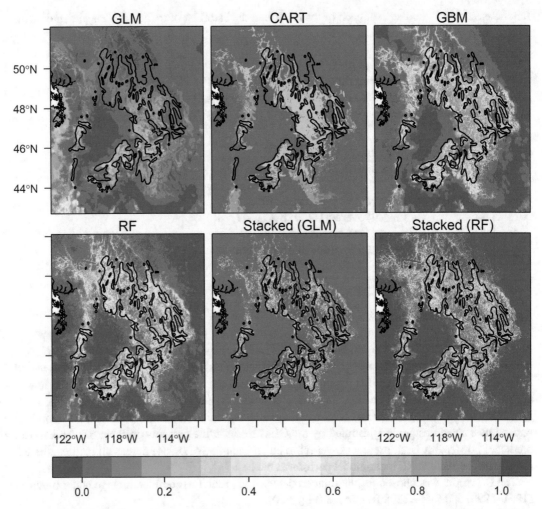

Fig. 5.10 Model predictions of the probability of western larch presence across its current range (black outline) in western North America using the rasterVis package. Models are: GLM, logistic regression; CART, classification and regression trees; GBM, boosted regression trees; RF, random forest; Stacked (GLM) linear ensemble of the four models; Stacked (RF), random forest ensemble of the four models

```
require(rasterVis)
require(lattice)
require(sp)

clim_ll <- raster::projectRaster(clim, crs = CRS("+init=epsg:4326"))
clim_sub_ll <- raster::subset(clim_ll,  x_vars)
clim_sub_ll <- raster::crop(clim_sub_ll, extent(spd_ll) * 1.3)

# Using raster package, make predictions to the raster stack of
#   climate variables. Predictions are probability of western
#   larch presence
xy_probs_ll <- lapply(names(mods_all),
                                 function(i) raster::predict(clim_sub_ll,
mods_all[[i]], type = 'prob', index = 1))
names(xy_probs_ll) <- names(mods_all)
```

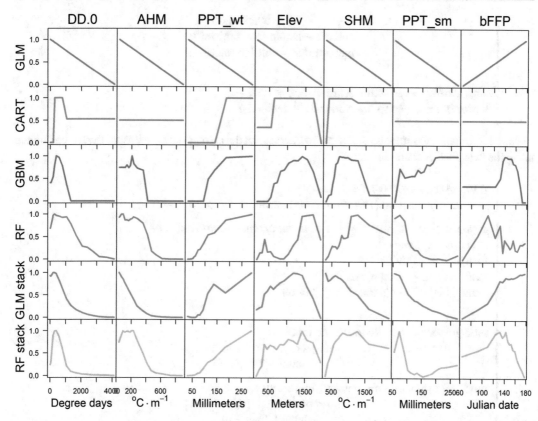

Fig. 5.11 Partial plots depicting the marginal effect of the predictor variables on predicting the geographic distribution of western larch across western North America. Marginal effect sizes were standardized to range between 0 and 1 for display purposes

```
# Convert list of predictions to a raster stack
xy_stack_ll <- stack(xy_probs_ll)
names(xy_stack_ll) <- names(xy_probs_ll)

# Reproject western larch distribution into lat-long projection
spd_ll <- spTransform(spd, CRSobj = CRS("+init=epsg:4326"))

# Reproject western North America boundary polygon to lat-long
bounds_ll <- spTransform(bounds, CRSobj = CRS("+init=epsg:4326"))

# Remove padding around the plot
lattice.options(
  layout.heights=list(bottom.padding=list(x=0),
                      top.padding=list(x=0)),
  layout.widths=list(left.padding=list(x=0),
                     right.padding=list(x=0)))

# Use levelplot to plot maps of predictions
plt <- rasterVis::levelplot(xy_stack_ll, margin=FALSE,
                  xlab = "", ylab = "",
                  names.attr = c("GLM", 'CART', "GBM",
                                "RF", "Stacked (GLM)",
```

```
                                    "Stacked (RF)"),
                        colorkey=list(space="bottom"),
                        par.settings = BuRdTheme())

plt +
latticeExtra:: layer(sp.lines(spd_ll, lwd = 1.5)) +
latticeExtra:: layer(sp.lines(bounds_ll, lwd = 1))
```

Variable importance measures also varied across models. Importance measures were calculated using the "varImp" function in "caret".

```
# VARIABLE IMPORTANCE
vl <- lapply(mods_all[1:4], function(x) varImp(x)$importance)

# Now merge the resultant variable importance data frames on
#   variable name
for(i in 1:length(vl)){
  vl[[i]]$vimp <- rownames(vl[[i]])
  names(vl[[i]]) <- c(names(vl)[i], 'vimp')
}

mod_importance <- Reduce(function(x, y)
                            merge(x, y, by = "vimp",
                                all.x = TRUE), vl)

mod_importance <- merge(mod_importance,
                        varImp(mods_all[[5]])[, 1, drop = F],
                        by.x = 'vimp', by.y = 0)
names(mod_importance)[ncol(mod_importance)] <- 'glm_stack'
tmp <- mod_importance$glm_stack

# Need to rescale the range of the glm_stack var importances as they don't run from
# 0-100 like the others
mod_importance$glm_stack <- (tmp - min(tmp)) / (max(tmp) -
                                min(tmp)) * 100
rownames(mod_importance) <- mod_importance$vimp
mod_importance <- mod_importance[, -1]
```

Table 5.5 Variable importance metrics for the final predictor variables for each of the machine learning algorithms tested[a]

Variable	Description	GLM	CART	GBM	RF	GLM stack
DD.0	Degree-days below 0° C (chilling degree days)	100.0	61.0	100.0	83.9	100.0
AHM	Annual heat moisture index	50.3	0.0	8.3	100.0	84.1
PPT_wt	Winter (Dec to Feb) precipitation (mm)	35.6	38.4	73.3	46.4	46.3
Elev	Elevation (m)	0.0	100.0	31.8	36.4	32.1
SHM	Summer heat moisture index	1.2	62.9	31.2	19.7	16.1
PPT_sm	Summer (Jun to Aug) precipitation (mm)	35.6	27.9	17.7	0.0	10.0
bFFP	Julian date on which the frost-free period begins	6.8	32.4	0.0	0.6	0.0

Degree days are calculated as the number of degrees that the mean daily temperature is below 0 °C over the course of the year

[a]Metrics are relativized to range between 0 and 100 using a linear transformation of the native scale for each algorithm.

```
mod_importance <- subset(mod_importance,
                         select = c(glm, rpart2, gbm, ranger,
                                    glm_stack))
mod_importance <- mod_importance[order(rowMeans(mod_importance),
                                       decreasing = T), ]
print(mod_importance)
```

The highest performing predictor for most models was the number of degree days below 0 °C (DD.0). This was not the case for the classification tree and RF models, in which elevation was the leading variable in the former and annual heat-moisture index (AHM) for the latter. Interestingly, though GBM and RF are both ensemble decision tree models, they disagreed on the importance of AHM, with GBM ranking this variable second to last. Given the order of predictor variable importance, it appears that predictors describing winter conditions rank somewhat higher than other seasonal variables, and temperature ranks higher than other climate variables.

Plotting the partial plots of each predictor helps identify the relationships between predictor and response variables, and differences in the relationships among models.

```
# Partial plots...
x_vars <- rownames(mod_importance)[order(rowMeans(mod_importance),
decreasing = T)]

require(pdp)
pp_list <- list()

# prediction function return logit for stacked models
pred_fun <- function(object, newdata){
  pr <- mean(predict(object = object, newdata = newdata, type = 'prob'))
  return(log(-(pr/(pr - 1))))
}

# Partial dependence plots with pdp package
for(j in 1:length(mods_all)){
  print(names(mods_all)[j])

  if(!names(mods_all)[j] %in% c('glm_stack', 'rf_stack'))
    pp_list[[j]] <- lapply(X = x_vars,
                FUN = function(z) partial(mods_all[[j]], pred.var = z, type = 'classification',
                                          which.class = 'yes', smooth = F, train = as.matrix
                                          (train[, -1]), prob = F))
  else
    pp_list[[j]] <- lapply(X = x_vars,
                FUN = function(z) partial(mods_all[[j]], pred.var = z, type = 'classification',
                                          which.class = 'yes', smooth = F, train = as.matrix
                                          (train[, -1]), pred.fun = pred_fun))
}

# Standardize the predicted partial plot values to between 0 - 100
#  for comparison. Each of the models will have a different range
#  of logit predictions
pp_std <- pp_list
for(i in 1:length(pp_std)){
```

```r
  for(j in 1:length(x_vars)){
    tmp <- pp_std[[i]][[j]]$yhat
    tmp <- (tmp - min(tmp)) / (max(tmp) - min(tmp))
    pp_std[[i]][[j]]$yhat <- tmp
  }
}

line_cols <- RColorBrewer::brewer.pal(n = 6, name = 'Dark2')

# PLOT THE PARTIAL PLOTS
#
par(mfrow = c(6, 7), omi = c(0.28, 0.52, 0.25, 0.1),
    mai = c(0.0, 0.0, 0.0, 0.0))


# All partial plots should have the same x-lim b/c they were
#   calculated off of the training data quantiles
for(i in 1:length(pp_std)){
  for(j in 1:length(x_vars)){

    plot(1, xlim = range(pp_std[[1]][[j]][, 1]), ylim = c(0, 1.2),
         type = 'n', xlab = "", ylab = "", yaxt = 'n', xaxt = 'n')

    if(i == 1)
      mtext(side = 3, adj = 0.5, text = x_vars[j], outer = F, cex = 1.2,
            line = 0.3)

    ifelse(i == 6,  axis(side = 1), axis(side = 1, labels = F))

    if(j == 1){
      axis(side = 2, at = seq(0, 1, 0.5), las = 1,
           labels = sprintf("%.1f", seq(0, 1, 0.5)))
      mtext(side = 2,
            c("GLM", 'CART', "GBM", "RF", "GLM stack",
            "RF stack")[i],
            line = 2.5, cex = 1.1)
    }

    tmp_x <- pp_std[[i]][[j]][, 1]
    tmp_y <- pp_std[[i]][[j]][, 2]

    if(all(is.nan(tmp_y)))
      tmp_y <- rep(0.5, length(tmp_y))

    lines(x = tmp_x, y = tmp_y, col = line_cols[[i]], lwd = 2)

    if(i == length(pp_std)){
      if(j == "DD.0")
        mtext(side = 1, adj = 0.5,
              text = "Degree days", line = 2.5)

      if(j %in% c("AHM", "SHM"))
        mtext(side = 1, adj = 0.5,
              text = quote("^{o}*C %.% m^-1), line = 2.5)

      if(j %in% c("PPT_wt", "PPT_sm"))
        mtext(side = 1, adj = 0.5,
```

```
                      text = "Millimeters", line = 2.5)

           if(j == "Elev")
             mtext(side = 1, adj = 0.5,
                      text = "Meters", line = 2.5)

          if(j == "bFFP")
            mtext(side = 1, adj = 0.5,
                      text = "Julian date", line = 2.5)
        }
      }

    }
```

The plots (Fig. 5.11) highlight differences among model flexibility and modeled relationships. The logistic regression (GLM) plots show the probability of western larch occurrence changes linearly across the range of the predictor variables. Decision trees (CART) exhibit step-like changes, adding some rigid non-linearity to the response curves. However, CART also failed to identify a relationship among three out of the seven predictor variables, which were the lowest importance in that model. The ensemble methods are capable of modeling smooth and flexible relationships.

With the exception of the logistic regression, the shape of the response curves (Fig. 5.11) generally exhibit similar patterns across the models. For DD.0, models suggest that western larch prefers a fairly narrow range of degree days[3] below 0 °C, with the probability of occurrence peaking around 500. RF models suggest a slightly larger window of tolerance for this variable compared to the other models. For the AHM variable, larger values indicate hot and dry conditions, while lower values indicate cool and moist climate. Western larch appears to prefer cool and moist climate, however, the range of optimal conditions varies across models with RF suggesting an optimal window between 0 and 500, while the stacked GLM meta-model shows an exponential decline in western larch tolerance to the annual heat-moisture index. For the summer heat-moisture index (SHM), however, modeled responses varied across models from positive asymptotic (CART, RF, stacked RF meta-model) to unimodal (GBM) to negative linear (stacked GLM meta-model). This variable exhibited fairly low importance across models, but shows variability among model inferences that can be drawn when only a single model is used. Probability of larch occurrence increased with increasing winter precipitation though some models suggest an asymptote around 200 mm (CART, GBM, and RF), while the remaining ensemble methods suggest a near linear relationship. Larch also appeared to occur within an elevation range between 500 and 1500 m, though models differed on the lower end of this range.

5.4.5   Model Prediction to Historical and Future Climates

After comparing model performance, spatial predictions, predictor variable importance, and response curves across models, the RF model was chosen to predict the western larch range under late twenty-first century predicted climate. Specifically, this model is used to predict if and how the climate envelope favored by western larch under recent climate conditions will shift geographically in the future under forecasted climate change conditions.

---

[3]Degree days are calculated as the number of degrees that the mean daily temperature is below 0°C over the course of the year.

Predictions to future climate conditions were made again using the raster package and were mapped using the "levelplot" function from the "rasterVis" package.

```
# FUTURE climate for western North America (CMIP3, A2 emissions,
#   2071 – 2100)
#  Downloaded  from:  http://www.cacpd.org.s3.amazonaws.com/ensemble_AOGCMs/
ENSEMBLE_A2_2080s_Bioclim_ASCII.zip
clim_A2_ll <- stack(list.files(paste0(inDir,
'/Data/climate/ENSEMBLE_A2_2080s_ASCII_LL'),
                        full.names = T,
                        pattern = "\\.tif$"))
# Digital elevation model downloaded from:
#http://www.cacpd.org.s3.amazonaws.com/climate_normals/Reference_files_ASCII.zip
dem_ll <- raster(paste0(inDir, "/Data/species_range_maps/elevation/NA/na_dem_ll.tif"))

# Need to do this to get the dem in the same extent as the climate
#  data
dem_ll <- projectRaster(dem_ll, to = clim_A2_ll)

clim_A2_ll$Elev <- dem_ll
names(clim_A2_ll)[c(6, 7, 19, 20, 22, 23, 27)] <- c('DD.0',
'DD.5', 'PPT_sm', 'PPT_wt', 'Tave_sm', 'Tave_wt', 'Elev')

# FUTURE climate stack across western North America including only
#  those variables in the models
clim_A2_sub_ll <- raster::subset(clim_A2_ll,  x_vars)

# FUTURE predicted probabilities using glmStack across western
#  North America
xy_A2_probs_rf_ll <- raster::predict(clim_A2_sub_ll,
                                     mods_all$ranger
                                     type = 'prob',
                                     index = 1)

# CURRENT predicted probabilities using glmStack across western
#  North America
clim_sub_ll2 <- raster::subset(clim_ll,  x_vars)

xy_probs_rf_ll <- raster::predict(clim_sub_ll2,
                                  mods_all$ranger
                                  type = 'prob', index = 1)
new_preds <- stack(xy_probs_rf_ll, xy_A2_probs_rf_ll)
names(new_preds) <- c('Current climate', 'Future climate (A2)')

new_preds_crop <- crop(new_preds, extent(c(-130, -105, 30, 60)))

spd_merc <- readOGR(list.files(paste0(inDir,
                        '/Data/species_range_maps/Little_range_maps/',
                    sp),
                             pattern = "\\.shp$",
                             full.names = T))
spd <- subset(spd_merc, subset = CODE == 1)
spd_ll <- spTransform(spd, CRSobj = CRS("+init=epsg:4326"))
```

```
lattice.options(layout.heights = list(bottom.padding =
                          list(x = 0),
                    top.padding = list(x = 0)),
                    layout.widths =
                          list(left.padding =
                                list(x = 0),
                          right.padding =
                                list(x = 0)))

plt <- rasterVis::levelplot(new_preds_crop, margin=FALSE,
                    xlab = "", ylab = "",
                    names.attr = c("Climate (1961 -
                                1990)",
                                "Future climate
                                (2080s)"),
                    colorkey=list(space="bottom"),
                    par.settings = BuRdTheme())

plt + latticeExtra:: layer(sp.lines(spd_ll, lwd = 1, col = 'black'))
```

Model predictions (Fig. 5.12) suggest a fairly significant northwesterly migration of the climate envelope favored by western larch by the end of the century. Favorable conditions for western larch all but disappear from its current southern extent and from the southern Cascade Mountains in Washington and Oregon. Within and adjacent to its current range, favorable climate for western larch appears to be pushing into higher elevations into the Rocky Mountains. The model predicts the range of western larch will shift well north into the Rocky Mountains and Northern Interior Mountains of British Columbia.

Caveats to this analysis:

1. Elevation is often a good predictor in ecological models, because it is a surrogate for joint climatic influences such as changes in temperature and precipitation regimes. However, often we are more interested in understanding the relationships between a response variable and the constituent climatic components that are coded for by elevation rather than elevation, per se. Furthermore, the climate at an elevation of, say, 1000 m at 30° latitude is much different than at 60° latitude, and therefore, models built with elevation as a predictor may not predict well outside of the training data.

2. Species respond to a variety of different environmental variables, not just climate (e.g., soils, competition from other species). The mapped predictions of the future distribution of western larch in Fig. 5.12 assume that climatic suitability is the only determinant of a species' range, and ignores other potential factors such as dispersal limitations and interspecific competition.

3. Also, tree species are fairly long lived, and experience a variety of climate conditions over generations. The climatic tolerances modeled using current climate may not fully represent their adaptive capacity to withstand climatic variability. The models suggest that favorable climate will likely shift north in the coming decades. Likewise, the climate may become unfavorable within the current distribution of western larch, possibly leading to range contraction in its southern extent.

4. Ecological responses often lag behind changes in climate, suggesting that the predicted migration of the climate envelope under predicted late twenty-first century climate conditions will not lead to immediate shifts in the distribution of western larch.

5. We used the most extreme GCM scenario (A2) for the decade 2080 for illustrative purposes to show potential changes in species distributions related to climate change. Projected changes in western larch distribution are therefore dependent upon the emissions assumptions of the GCM and the fidelity of future climate predictions.

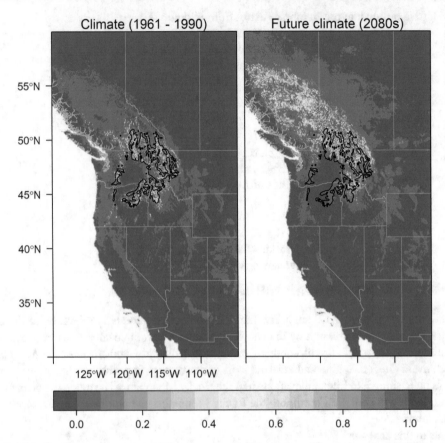

Fig. 5.12 Model predictions of the probability of western larch presence across western North America under past climate conditions (left panel), and a future predicted climate using the ensemble A2 high emissions scenario for 2071–2100 (right panel). Predictions were made using the GLM Stack model fit to the 1961–1990 climate data within the "caret" package

5.4.6   Further Reading

The field of machine learning, and more broadly, that of data science, is a burgeoning field that is quickly becoming a staple of scientific research. This chapter was merely a brief introduction to some of the key concepts and applications of the science. Some invaluable resources to help further the introduction to basic machine learning methods include the *Introduction to Statistical Learning* (James et al. 2013), which greatly expands on the themes presented here for a broader audience. Similarly, *Statistical Learning from a Regression Perspective* (Berk 2008) presents machine learning, using concepts familiar for those who have used linear and generalized linear models in the past. First editions of both texts are available for free through Google Scholar. Another classic and more advanced text, *The Elements of Statistical Learning* (Friedman et al. 2001), is widely cited and is appropriate for more novice machine learning users. Several journal articles have also been written that are geared towards the ecological community and meant as introduction resources (Elith et al. 2006, Cutler et al. 2007, Olden et al. 2008, Franklin 2010, Thessen 2016).

Appendix 1

Species	Life.form
Agrostis canina L.	H
Anemone nemorosa L.	G
Antennaria dioica (L.) Gaertn.	H
Anthoxanthum odoratum L.	H
Arrhenatherium elatius (L.) J. &. C. Presl.	H
Asperula cynanchica L.	H
Betula pendula Roth	P
Briza media L.	H
Campanula patula L.	T
Campanula sparsa Friv.	T
Carpinus betulus L.	P
Chamaecytisus austriacus (L.) Link	Ch
Corylus avellana L.	P
Crataegus monogyna Jacq.	P
Dryopteris filix-mas (L.) Schott	G
Epilobium montanum L.	H
Euphorbia amygdaloides L.	H
Fagus moesiaca (K. Maly) Czecz.	P
Festuca ovina L.	H
Fragaria vesca L.	H
Galium odoratum (L.) Scop.	H
Genista depressa M.Bieb.	Ch
Genista sagittalis L.	Ch
Gentiana asclepiadea L.	H
Helianthemum nummularium (L.) Mill.	Ch
Hieracium acuminatum Jord.	H
Hieracium piloselloides Vill.	H
Holcus lanatus L.	H
Hordelymus europaeus (L.) Jess. ex Harz	H
Jasione montana L.	H
Lotus corniculatus L.	H
Luzula forsteri (Sm.) DC.	H
Luzula luzuloides (Lam.) Dandy & Wilmott	H
Malus sylvestris (L.) Mill.	P
Melica uniflora Retz.	H
Nardus stricta L.	H
Orobanche gracilis Sm.	G

© Springer Nature Switzerland AG 2020
M. Lakicevic et al., *Introduction to R for Terrestrial Ecology*,
https://doi.org/10.1007/978-3-030-27603-4

Species	Life.form
Pastinaca hirsuta Panč.	H
Poa nemoralis L.	H
Populus tremula L.	P
Prunella laciniata (L.)	H
Prunella vulgaris L.	H
Pulmonaria officinalis L.	H
Quercus petraea (Matt.) Liebl.	P
Quercus pubescens Willd.	P
Rosa tomentosa Sm.	T
Rosa vosagiaca N.H.F. Desp.	P
Salix fragilis L.	P
Sanguisorba minor Scop.	H
Saxifraga rotundifolia L.	H
Scabiosa columbaria L.	H
Senecio viscosus L.	T
Senecium nemorensis L.	H
Solanum hispidum Pers.	T
Sorbus aucuparia L.	P
Sorbus torminalis (L.) Crantz	P
Teucrium chamaedrys L.	Ch
Thymus pulegioides L.	Ch
Trifolium alpestre L.	H
Trifolium pratense L.	H
Vaccinium myrtillus L.	Ch
Veronica chamaedrys L.	Ch
Viola sylvestris Lam.	H

Appendix 2

Species	Height	Damages	Description
Platanus x acerifolia (Aiton) Willd.	85,3	F	Protected
Ailanthus altissima (P. Mill.) Swingle	32,8	F	Invasive
Fagus moesiaca (K. Maly) Czecz.	49,2	F	Protected
Ailanthus altissima (P. Mill.) Swingle	82,1	F	Invasive
Fagus moesiaca (P. Mill.) Swingle	45,9	T	Protected
Carpinus betulus L.	49,2	F	Regular
Quercus robur L.	82,1	F	Protected
Ailanthus altissima (P. Mill.) Swingle	36,1	F	Invasive
Acer negundo L.	26,3	F	Invasive
Platanus x acerifolia (Aiton) Willd.	82,0	F	Protected
Ailanthus altissima (P. Mill.) Swingle	98,4	F	Invasive
Populus alba L.	82,0	F	Regular
Populus alba L.	68,9	F	Regular
Populus alba L.	82,1	F	Regular
Robinia pseudoacacia L.	36,1	T	Invasive
Robinia pseudoacacia L.	42,7	F	Invasive
Robinia pseudoacacia L.	49,2	F	Invasive
Populus alba L.	91,9	F	Regular
Acer negundo L.	34,5	T	Invasive
Acer negundo L.	37,7	T	Invasive
Quercus robur L.	65,6	F	Protected
Ailanthus altissima (P. Mill.) Swingle	36,1	F	Invasive
Ailanthus altissima (P. Mill.) Swingle	49,2	F	Invasive
Ailanthus altissima (P. Mill.) Swingle	41,1	T	Invasive
Ailanthus altissima (P. Mill.) Swingle	29,5	F	Invasive
Carpinus betulus L.	42,7	T	Regular
Fagus moesiaca (K. Maly) Czecz.	49,2	T	Protected
Fagus moesiaca (K. Maly) Czecz.	29,5	F	Protected
Carpinus betulus L.	30,0	T	Regular
Carpinus betulus L.	36,1	F	Regular
Platanus x acerifolia (Aiton) Willd.	82,0	F	Protected
Platanus x acerifolia (Aiton) Willd.	82,0	F	Protected
Platanus x acerifolia (Aiton) Willd.	55,8	T	Protected
Platanus x acerifolia (Aiton) Willd.	80,5	F	Protected
Platanus x acerifolia (Aiton) Willd.	72,2	T	Protected
Quercus robur L.	75,5	F	Protected
Fagus moesiaca (K. Maly) Czecz.	31,1	F	Protected

© Springer Nature Switzerland AG 2020
M. Lakicevic et al., *Introduction to R for Terrestrial Ecology*,
https://doi.org/10.1007/978-3-030-27603-4

Species	Height	Damages	Description
Robinia pseudoacacia L.	42,8	T	Invasive
Robinia pseudoacacia L.	28,9	F	Invasive
Robinia pseudoacacia L.	41,0	F	Invasive
Fagus moesiaca (K. Maly) Czecz.	54,1	F	Protected
Fagus moesiaca (K. Maly) Czecz.	50,8	F	Protected
Tilia argentea DC.	65,2	T	Invasive
Tilia argentea DC.	60,4	T	Invasive
Tilia argentea DC.	55,3	T	Invasive
Tilia argentea DC.	58,4	F	Invasive
Tilia argentea DC.	52,7	T	Invasive
Tilia argentea DC.	60,2	F	Invasive
Tilia argentea DC.	54,3	T	Invasive
Betula pendula Roth	45,5	F	Regular
Betula pendula Roth	50,2	F	Regular
Betula pendula Roth	53,4	F	Regular
Betula pendula Roth	56,3	F	Regular
Betula pendula Roth	52,4	F	Regular
Betula pendula Roth	58,3	F	Regular
Betula pendula Roth	49,7	F	Regular

Variable	Description
Tmin	Minimum temperature for a given month (°C)**
Tmax	Maximum temperature for a given month (°C)**
PPT	Total precipitation for a given month (mm)
MAT	Mean annual temperature (°C)**
MCMT	Mean temperature of the coldest month (°C)**
MWMT	Mean temperature of the warmest Month (°C)**
TD	Difference between MCMT and MWMT, as a measure of continentality (°C)**
MAP	Mean annual precipitation (mm)
MSP	Mean summer (May to Sep) precipitation (mm)
AHM	Annual heat moisture index, calculated as (MAT+10)/(MAP/1000)**
SHM	Summer heat moisture index, calculated as MWMT/(MSP/1000)**
DD.0	Degree-days below 0 °C (chilling degree days)
DD.5	Degree-days above 5 °C (growing degree days)
NFFD	Number of frost-free days
bFFP	Julian date on which the frost-free period begins
eFFP	Julian date on which the frost-free period ends
PAS	Precipitation as snow (mm)
EMT	Extreme minimum temperature over 30 years (°C)**
Eref	Hargreave's reference evaporation
CMD	Hargreave's climatic moisture index
CMI	Hogg's climate moisture index
cmiJJA	Hogg's summer (June–Aug) climate moisture index
Tave_wt	Winter (Dec–Feb) mean temperature (°C)**
Tave_sm	Summer (June–Aug) mean temperature (°C)**
PPT_wt	Winter (Dec–Feb) precipitation (mm)
PPT_sm	Summer (June–Aug) precipitation (mm)

**Temperature variables are giving in units of degrees Celsius $* 10$

List of the R Packages Used in the Book (Alphabetical Order)

Package name	Citation
"agricolae"	Felipe de Mendiburu (2019). agricolae: Statistical Procedures for Agricultural Research. R package version 1.3-1. https://CRAN.R-project.org/package=agricolae
"car"	John Fox and Sanford Weisberg (2011). An {R} Companion to Applied Regression, Second Edition. Thousand Oaks CA: Sage. https://socserv.socsci.mcmaster.ca/jfox/Books/Companion
"caret"	Max Kuhn. Contributions from Jed Wing, Steve Weston, Andre Williams, Chris Keefer, Allan Engelhardt, Tony Cooper, Zachary Mayer, Brenton Kenkel, the R Core Team, Michael Benesty, Reynald Lescarbeau, Andrew Ziem, Luca Scrucca, Yuan Tang, Can Candan and Tyler Hunt. (2019). caret: Classification and Regression Training. R package version 6.0-84. https://CRAN.R-project.org/package=caret
"caretEnsemble"	Zachary A. Deane-Mayer and Jared E. Knowles (2016). caretEnsemble: Ensembles of Caret Models. R package version 2.0.0. https://CRAN.R-project.org/package=caretEnsemble
"datasets"	R Core Team (2019). R: A language and environment for statistical computing. R Foundation for Statistical Computing, Vienna, Austria. https://www.R-project.org/
"devtools"	Hadley Wickham, Jim Hester and Winston Chang (2018). devtools: Tools to Make Developing R Packages Easier. R package version 2.0.1. https://CRAN.R-project.org/package=devtools
"doParallel"	Microsoft Corporation and Steve Weston (2018). doParallel: Foreach Parallel Adaptor for the 'parallel' Package. R package version 1.0.14. https://CRAN.R-project.org/package=doParallel
"dplyr"	Hadley Wickham, Romain François, Lionel Henry and Kirill Müller (2018). dplyr: A Grammar of Data Manipulation. R package version 0.7.8. https://CRAN.R-project.org/package=dplyr
"emmeans"	Russell Lenth (2019). emmeans: Estimated Marginal Means, aka Least-Squares Means. R package version 1.3.3. https://CRAN.R-project.org/package=emmeans
"gbm"	Brandon Greenwell, Bradley Boehmke, Jay Cunningham and GBM Developers (2019). gbm: Generalized Boosted Regression Models. R package version 2.1.5. https://CRAN.R-project.org/package=gbm
"ggplot2"	H. Wickham. ggplot2: Elegant Graphics for Data Analysis. Springer-Verlag New York, 2016.
"ggpolypath"	Michael D. Sumner (2016). ggpolypath: Polygons with Holes for the Grammar of Graphics. R package version 0.1.0. https://CRAN.R-project.org/package=ggpolypath

© Springer Nature Switzerland AG 2020
M. Lakicevic et al., *Introduction to R for Terrestrial Ecology*,
https://doi.org/10.1007/978-3-030-27603-4

Package name	Citation
"latticeExtra"	Sarkar, Deepayan and Andrews, Felix (2016) latticeExtra: Extra Graphical Utilities Based on Lattice. R package version 0.6-28. https://CRAN.R-project.org/package=latticeExtra
"multicompView"	Russell Lenth (2019). emmeans: Estimated Marginal Means, aka Least-Squares Means. R package version 1.3.3. https://CRAN.R-project.org/package=emmeans
"pdp"	Greenwell, Brandon M. (2017) pdp: An R Package for Constructing Partial Dependence Plots. The R Journal, 9(1), 421–436. https://journal.r-project.org/archive/2017/RJ-2017-016/index.html
"plyr"	Hadley Wickham (2011). The Split-Apply-Combine Strategy for Data Analysis. Journal of Statistical Software, 40(1), 1-29. http://www.jstatsoft.org/v40/i01/
"ranger"	Marvin N. Wright, Andreas Ziegler (2017). ranger: A Fast Implementation of Random Forests for High Dimensional Data in C++ and R. Journal of Statistical Software, 77(1), 1–17. doi:10.18637/jss.v077.i01
"rasterVis"	Oscar Perpinan Lamigueiro and Robert Hijmans (2018), rasterVis. R package version 0.45.
"reshape"	H. Wickham. Reshaping data with the reshape package. Journal of Statistical Software, 21(12), 2007.
"rgbif"	Chamberlain S, Barve V, Mcglinn D, Oldoni D, Desmet P, Geffert L, Ram K (2019)._rgbif: Interface to the Global Biodiversity Information Facility API_. R package version 1.3.0. https://CRAN.R-project.org/package=rgbif>
"Rmisc"	Ryan M. Hope (2013). Rmisc: Rmisc: Ryan Miscellaneous. R package version 1.5. https://CRAN.R-project.org/package=Rmisc
"rpart"	Terry Therneau and Beth Atkinson (2018). rpart: Recursive Partitioning and Regression Trees. R package version 4.1-13. https://CRAN.R-project.org/package=rpart
"sp"	Pebesma, E.J., R.S. Bivand, 2005. Classes and methods for spatial data in R. R News 5(2). https://cran.r-project.org/doc/Rnews/.
"stats"	R Core Team (2019). R: A language and environment for statistical computing. R Foundation for Statistical Computing, Vienna, Austria. https://www.R-project.org/
"usmap"	Paolo Di Lorenzo (2018). usmap: US Maps Including Alaska and Hawaii. R package version 0.4.0. https://CRAN.R-project.org/package=usmap
"vegan"	Jari Oksanen, F. Guillaume Blanchet, Michael Friendly, Roeland Kindt, Pierre Legendre, Dan McGlinn, Peter R. Minchin, R. B. O'Hara, Gavin L. Simpson, Peter Solymos, M. Henry H. Stevens, Eduard Szoecs and Helene Wagner (2018). vegan: Community Ecology Package. R package version 2.5-3. https://CRAN.R-project.org/package=vegan

References

Adler, J. 2012. *R in a nutshell: A desktop Quick Reference*. Sebastopol: O'Reilly Media, Inc.

Begon, M., J. L. Harper, and C. R. Townsend. 1996. *Ecology: Individuals, Populations, and Communities*. 3rd edition. Cambridge: Blackwell Science Ltd.

Berk, R. A. 2008. *Statistical learning from a regression perspective*. Springer.

Breiman, L. 1996. *Stacked regressions*. Machine Learning 24:49–64.

Breiman, L. 2001a. *Statistical modeling: The two cultures (with comments and a rejoinder by the author)*. Statistical science 16:199–231.

Breiman, L. 2001b. *Random forests*. Machine learning 45:5–32.

Breiman, L. 2017. *Classification and regression trees*. Routledge.

Brekke, L., B. L. Thrasher, E. P. Maurer, and T. Pruitt. 2013. *Downscaled CMIP3 and CMIP5 climate projections. Release of Downscaled CMIP5 Climate Projections, Comparison with Preceding Information, and Summary of User Needs*.

Cutler, D. R., T. C. Edwards Jr, K. H. Beard, A. Cutler, K. T. Hess, J. Gibson, and J. J. Lawler. 2007. *Random forests for classification in ecology*. Ecology 88:2783–2792.

De'ath, G., and K. E. Fabricius. 2000. *Classification and regression trees: a powerful yet simple technique for ecological data analysis*. Ecology 81:3178–3192.

Elith, J., and J. Leathwick. 2017. *Boosted Regression Trees for ecological modeling. R documentation*. Available at https://cran. r-project. org/web/packages/dismo/vignettes/brt. pdf.

Elith, J., C. H. Graham, R. P. Anderson, M. Dudík, S. Ferrier, A. Guisan, R. J. Hijmans, F. Huettmann, J. R. Leathwick, A. Lehmann, and others. 2006. *Novel methods improve prediction of species' distributions from occurrence data*. Ecography 29:129–151.

Elith, J., J. R. Leathwick, and T. Hastie. 2008. *A working guide to boosted regression trees*. Journal of Animal Ecology 77:802–813.

Fedor, P., and M. Zvaríková. 2019. *Biodiversity Indices*. In: Encyclopedia of Ecology (Second Edition), Fath, B.D. (Ed). Oxford: Elsevier Science Ltd, pp. 337-346.

Franklin, J. 2010. *Mapping species distributions: spatial inference and prediction*. Cambridge University Press.

Friedman, J., T. Hastie, and R. Tibshirani. 2001. *The elements of statistical learning*. Springer series in statistics New York.

Good, P.I. 2005. *Permutation, Parametric, and Bootstrap Tests of Hypotheses*. New York: Springer-Verlag New York.

Hamann, A., T. Wang, D. L. Spittlehouse, and T. Q. Murdock. 2013. *A comprehensive, high-resolution database of historical and projected climate surfaces for western North America*. Bulletin of the American Meteorological Society 94:1307–1309.

James, G., D. Witten, T. Hastie, and R. Tibshirani. 2013. *An introduction to statistical learning*. Springer.

Kéry, M. 2010. *Introduction to WinBUGS for Ecologists*. Burlington: Academic Press.

Kuhn, M. 2012. *Variable selection using the caret package*. URL http://cran. cermin. lipi. go. id/web/packages/caret/vignettes/caretSelection. pdf.

Kuhn, M. 2018. caret: *Classification and Regression Training*.

Lalanne, C., M. Mesbah. 2016. *Biostatistics and Computer-based Analysis of Health Data Using SAS*. London: Elsevier Science Ltd.

Little Jr, E. 2006. *Digital representations of tree species range maps from 'Atlas of United States Trees.'*

Ma, Z.S. 2018. *Measuring Microbiome Diversity and Similarity with Hill Numbers*. In: Metagenomics: Perspectives, Methods and Applications, Nagarajan, M. (Ed). London: Academic Press, pp. 157-178.

Magurran, A.E. (2004) *Measuring Biological Diversity*. London: Blackwell Publishing.

© Springer Nature Switzerland AG 2020

M. Lakicevic et al., *Introduction to R for Terrestrial Ecology*,

https://doi.org/10.1007/978-3-030-27603-4

Meehl, G. A., C. Covey, T. Delworth, M. Latif, B. McAvaney, J. F. Mitchell, R. J. Stouffer, and K. E. Taylor. 2007. *The WCRP CMIP3 multimodel dataset: A new era in climate change research*. Bulletin of the American Meteorological Society 88:1383–1394.

Miller, H. J. 2004. *Tobler's first law and spatial analysis*. Annals of the Association of American Geographers 94:284–289.

National Park Service 2016. *The National Parks: Index 2012-2016*. Washington, D.C.: U.S. Department of the Interior.

Olden, J. D., J. J. Lawler, and N. L. Poff. 2008. *Machine learning methods without tears: a primer for ecologists*. The Quarterly review of biology 83:171–193.

Petrie, A., and C. Sabin. 2005. *Medical statistics at a glance*. Malden: Blackwell Publishing, Inc.

Rakić, T., M. Lazarević, Ž. S. Jovanović, S. Radović, S. Siljak-Yakovlev, B. Stevanović, and V. Stevanović. 2014. *Resurrection plants of the genus Ramonda: prospective survival strategies – unlock further capacity of adaptation, or embark on the path of evolution?* Front. Plant Sci. 4:550.

Raunkiaer, C. 1934. *The life forms of plants and statistical geographical plant geography*. London: Oxford University Press.

Smalheiser, N.R. 2017. *Data Literacy*. London: Academic Press.

Thessen, A. 2016. Adoption of Machine Learning Techniques in Ecology and Earth Science. One Ecosystem 1:e8621.

Vuković, N. 1997. *PC verovatnoća i statistika*. Beograd: Fakultet organizacionih nauka.

Wolpert, D. H. 1992. *Stacked generalization*. Neural Networks 5:241–259.

Printed in the United States
by Baker & Taylor Publisher Services